Illustrated

AF271319

CASE
Tractor
B U Y E R' S ★ G U I D E™

Peter Letourneau

Motorbooks International
Publishers & Wholesalers

For over 150 years, J. I. Case Company and its dealers have served world agriculture by manufacturing and supplying the highest quality of farm equipment. As a former Case employee and dealer, I take great pride in my past association with the company. I dedicate this book to all those with whom I worked in my years at Case. It was great!

First published in 1993 by Motorbooks International Publishers & Wholesalers, PO Box 2, 729 Prospect Avenue, Osceola, WI 54020 USA

© Peter Letourneau, 1993

Motorbooks International is a certified trademark, registered with the United States Patent Office

The information in this book is true and complete to the best of our knowledge. All recommendations are made without any guarantee on the part of the author or Publisher, who also disclaim any liability incurred in connection with the use of this data or specific details

We recognize that some words, model names and designations, for example, mentioned herein are the property of the trademark holder. We use them for identification purposes only. This is not an official publication

Motorbooks International books are also available at discounts in bulk quantity for industrial or sales-promotional use. For details write to Special Sales Manager at the Publisher's address

Library of Congress Cataloging-in-Publication Data

Letourneau, Peter A.
    Illustrated case tractor buyer's guide / Peter Letourneau.
        p.   cm.—(Motorbooks International illustrated buyer's guide)
    ISBN 0-87938-784-X
    1. Case tractors—Purchasing.   2. Case tractors—History.   3. Case tractors—Collectors and collecting.   I. Title.   II. Series.
TL233.5.L47   1993
629.224—dc20   93-19414

On the front cover: The classic 1952 Case DCS Sugar Cane Special owned by Tom Graverson of Indiana. *Andrew Morland*

Printed and bound in the United States of America

Contents

Acknowledgments

My Case cap is off to all those who aided me in creating this book. In particular, I wish to thank: Rick James, Dave Rogers, and Eldon Brumbaugh at J. I. Case Company; Noel Nelson of Hawley, Minnesota, who was particularly generous with information related to his family's experiences as Case dealers and collectors; the Western Minnesota Steam Thresher Reunion, for the 1992 Case 150th anniversary celebration; John Skarstad, Curator of the Shields Library, Special Collections, University of California, Davis, and the staff of UC Davis Illustration Services; and Michael Dregni at Motorbooks International, whose attention and encouragement is appreciated.

Introduction and Investment Ratings

Introduction and Investment Ratings

The Illustrated Case Tractor Buyer's Guide is a reference to Case two- and four-cylinder gasoline-, kerosene-, and LP-Gas-fueled tractors built for the model years 1912 through 1955. It was created to help collectors and potential collectors identify, evaluate, and select a Case gas tractor from this era. To that end, it reviews the significant features of each model; summarizes their specifications; presents production and serial number information; and offers general and specific buying tips.

The guide also features a simplified five-star rating system to assist the enthusiast in determining the investment potential of these tractors. The number of stars assigned to a particular model is based on its uniqueness, availability, and desirability, and is translated as follows:

★★★★★ Five stars—the best investment. Tractors with a five-star rating are expensive, and continued appreciation can be expected. Such tractors are often sold or traded between major collectors without the general collecting public being aware of their availability.

★★★★ Four stars—excellent investment potential. Four-star rated tractors also tend to be expensive and are likely to appreciate at a rate above that of inflation. As with five-star tractors, four-star tractors are often traded among established collectors who pay close attention to the market.

★★★ Three stars—very good investment, if ownership satisfaction is considered. Values should keep pace with inflation, for a three-star tractor in good to very good condition. An unrestored three-star tractor, bought at a below market price, presents a good opportunity for return.

★★ Two stars—good investment, but with cost of ownership. High-production (common) older models fall in the two-star category, as do most newer models.

★ One star—marginal or poor investment potential. Tractors in which there is little interest and for which routine restoration costs would likely exceed any potential medium- to long-range return.

When considering these ratings, keep in mind that the market for both unrestored and restored tractors is a dynamic one. As the more desirable Case models are bought up, and as more people begin or add to their collections, the values of many less-desirable models will increase at a rate beyond that of today.

While the five-star rating system is a prominent feature of this book, it is offered solely as a guideline to each model's present investment potential. It is not our intention to rob anyone of their enthusiasm for any given model of Case tractor. People collect or restore tractors for many reasons, and return on investment is not necessarily first among them. Your own judgment and sentiments will certainly play a large part in such decisions.

For further guidance, we would encourage you to join the J. I. Case Collector's Association, a first-rate organization whose quarterly publication *Old Abe's News* is among the finest enthusiast magazines available. Every issue features historical and technical information, as well as a classified section through which members buy and sell tractors. For membership information write, J. I. Case Collectors Association Inc., c/o Crooked Hollow Farm Publishing, 4004 Coal Valley Road, Vinton, Ohio 45686–9741.

Also recommended is the J.I. Case Heritage Foundation, Box 5128, Bella Vista, Arkansas 72714-0128. This group's quarterly publication is titled *The Heritage Eagle*.

Chapter 1

Buying a Collector Tractor

Once you locate a tractor that interests you, put your excitement on hold and take an objective look at what you're getting into. It's impossible to make an intelligent decision as to a particular tractor's condition or potential without first inspecting it thoroughly.

Robert N. Pripps, author of the *Illustrated John Deere Two-Cylinder Tractor Buyer's Guide* and *How to Restore Your Farm Tractor* (both published by Motorbooks International) has developed a Buyer's Check List that every enthusiast should adopt as his or her own.

A straightforward, commonsense tool, Pripps's check list (with one or two additions of my own) and his comments concerning its use are reproduced below. Once you've used his list, I'm certain you will agree that Mr. Pripps deserves every tractor collector's thanks.

Buyer's Check List

"For best results, take this check list with you when you go to evaluate a tractor for purchase. Check the items off as they appear on the list, making notes on each section as you go. The purpose of this is twofold: first, it's an orderly way to complete the evaluation with as much rationality as possible, and second, it gives you a good record of the evaluation for comparison with those of other tractors you find. The check list will be followed by explanations of the items and what to look for.

General Appearance
Sheet metal, grille, and fenders
Tires and wheels
Steering wheel
Exhaust
Oil, water, and fuel system (for leaks)
Model designation
Serial number

Steering
Steering wheel free play
Kingpin free play
Radius rod free play
Front wheel bearing free play
Drag arm or arms free play

Engine
Evidence of crack repair in block and head
Condition of manifolds
Oil in crankcase
Filter in place
Water in crankcase
Oil in water
Belts
Hoses
Radiator cap
Air cleaner
Carburetor controls
Fuel tank(s) and fuel in tank(s)
Fuel filter, sediment bowl, and shutoff (to see if open)

Electrical
Magneto
Battery in place, condition, and water
Cables and terminals
Generator and brushes

Starter—visible condition
Key and switch—location and operation
Ammeter indication—key on and off
Fuel, temperature, and other gauges
Plug wire condition
General wiring condition
Lights

Clutch and Transmission
Clutch operation
Shift lever operation
Oil level
Water in oil
Leaks, welds, and repairs

Rear Axle
Housing cracks, repairs, and leaks
Lubricant level and water in lubricant
Axle wheel seals
Brakes and linings and linkage

After starting:
Engine
Oil pressure
Smoke—exhaust pipe and breather
Knocking
Missing
Temperature stabilization
Throttle response—revolutions per minute
    (rpm) range and governor operation
Oil, water, and hydraulic leaks
Starter operation
Generator charging

Clutch and Transmission
Clutch releasing completely
Gear selection
Clutch engaging smoothly
Clutch slippage
Free pedal
PTO operation

Brakes
Left and right brake power

Hydraulic System
Lift ability
Leak-down
Smoothness

Road-Field Test
Steering shimmy and binding
Brakes
Engine operation under load

Hydraulics operation
Water temperature
Inappropriate noises

"It is important that the 'before starting' test be done before attempting to start the engine, unless the tractor has been recently run. Not only will this prevent damage from things like lack of oil, but it will allow you to check for water in the oil, or oil in the water, before operation gets everything mixed up. It will, in addition, serve as a setup check, so you don't, for example, attempt to start the engine with the fuel shut off.

"The general appearance items are self-explanatory, except for model designation and serial number. These are included so that proper credit will be given if the tractor is historically significant or rare. Lack of definite evidence of model designation or serial number can impede the acquisition of parts and may indicate that the configuration is not original. Generally, the serial number is the only 'hard data' on the bill of sale, by which to identify the tractor.

"Another important item under this heading is tires. With a new set of current-production tires costing between $600 and $1,000, their condition vitally affects the value of the tractor. To obtain good tires with tread appropriate to the year of the tractor can be difficult or even impossible, and costly. . . .

"Steering wheel free play of 2in, measured at the rim, is about the limit of acceptability for more-or-less modern tractors. Obviously . . . steering . . . will be more sloppy . . . for models lacking provision for wear adjustment.

"As you make the engine checks, make the engine ready to start—that is, open the fuel shutoff, add oil and water as necessary, and so forth—if the engine is operable. If the engine is inoperable, do your best to determine why. Is compression developing? Is spark occurring? Is the fuel getting through? Is the engine 'stuck?' A truly stuck engine can be a disaster, but [an engine] can often be 'unstuck' by liberally applying penetrating oil, or WD–40, and then slowly towing [the tractor] in top gear and gingerly engaging the clutch. If the tires slide, let the oil soak for another day or two. Many unstuck [engines] operate perfectly after undergoing this procedure.

"The ammeter indication item is intended to show that the ignition system and switch are functional, by having you observe an indication of 'discharge' when the switch is turned on.

"It is important to ascertain that the clutch is actually released when the lever or pedal is in the release position and that the transmission is actually in neutral before you attempt to manually start the engine. Failure to do so could result in your being run over!

"It's best to check the hydraulics with a heavy implement such as a plow or mower deck. The system should be able to raise and hold anything designed for it with ample reserve. With the engine off, the system should not let the implement down for at least ten minutes.

"Ideally, you can operate the tractor in a field with an implement. . . . You should also take it where you can operate it at top road speed. Try all the gears and the brakes. Operate it long enough for things to get warm. Listen for any unusual sounds as it warms up.

"Remember, with all these tests, the main things you are to determine are these: is the price fair, and do you want to get involved in the amount of work required to put the tractor into the shape you desire?"

Bill of Sale

Once you make your deal, obtain a bill of sale. Include the seller's name and address, your name and address, price, and a full description of the tractor including serial number or other identification. Have the seller sign it and, if you would feel more comfortable, insist that it be notarized.

The famous and popular Case eagle trademark featured Old Abe perched on a globe. It was adopted by the company in 1894 and used until 1969. *Case Company*

Chapter 2

The Paterson/Case Experiment 1892

Rating	Model	Remarks
★★★★★	Paterson/Case	Experimental model

In 1992, J. I. Case Company celebrated the 150th anniversary of its founding. In recognition of this milestone, Case equipment was prominently displayed at antique tractor and threshing events across North America.

Built in 1869, Steam Engine No. 1 was the first Case portable steam engine. It developed 8–10hp. *Case Company*

J. I. Case Company's history as a supplier of farm power is almost as old as the company itself. By the late 1860s, steam had become a significant source of horsepower, particularly to threshing machine operators. Case was, of course, a leading manufacturer of threshing machines. Because customers for threshing machines were also potential buyers of steam engines, it made sense for Case to build and offer its own steam engines.

In 1869, the company led by Jerome Increase Case built its first portable steam engine, a unit that developed 8 to 10hp. Case went on to build some 36,000 stationary, portable, and traction steam engines, more than any other manufacturer of farm equipment. The last steam traction unit was built in 1924, fifty-five years after Steam Engine No. 1 was hitched to a team of horses and pulled from the Racine, Wisconsin, shops.

The year 1992 was significant for another reason as well, as it also marked the 100th anniversary of the first Case gasoline tractor. An experimental model, it was one of the first practical gasoline tractors.

The following description of the tractor appeared in the Case Centennial issue of *Farm Implement News*, dated January 8, 1942: the "experimental gas tractor . . . was moved from factory to a farm twelve miles distant for and where it was used in threshing. A great deal of the work on this machine was done by David (Pryce) Davies, then a young draftsman working in the experimental department. . . . Concerning this tractor Mr. Davies writes: 'the engine was of the four-cycle type and with two cylinders. A working stroke was obtained every revolution of the crankshaft. Very little was known at this early date regarding either carburetion or ignition and there was not a single manufacturer of carburetors or apparatus pertaining to ignition. The chassis of the tractor, with the exception of the main frame, consisted largely of parts such as were in use at that time on our steam tractor. One forward speed only was provided, the reverse being accomplished by sliding a key which could be shifted into neutral position or into forward or reverse speeds. One of the gears driven by the key was operated direct from the pinion on the crankshaft, and the other by means of an idler interposed between it and the pinion on the crankshaft. This was to secure the opposite motion for reversing the tractor. Because of the lack of proper carburetor and ignition,

it was decided to drop the tractor at that time.'"

According to R. B. Gray, author of *The Agricultural Tractor 1855–1950*, the tractor's engine developed between 16hp and 20hp. He noted that its crude "make-and-break" ignition system featured a bolt in the piston head, which made contact with a stationary insulated electrode on the compression stroke. Spark occurred just after dead center on the outward stroke.

The tractor is sometimes referred to as the "Paterson" tractor, as its engine was designed by William Paterson. There is evidence that suggests that the entire machine was built by Paterson under contract to Case.

Little more is known of the Paterson/Case tractor. There is no record of what became of it. It was a notable experiment, but apparently not an encouraging one, as two decades passed before Case introduced its first production gasoline tractor.

Built in 1892, the first Case prototype gasoline tractor featured a Paterson "Balanced Gas Engine." It was one of the first practical gas tractor designs. *Case Company*

Chapter 3

The First Case Gas Tractors 1912–1916

Models 30–60, 20–40, 12–25, and 10–20

Rating	Model	Remarks
★★★★★	30–60	First production Case tractor. 500 built; five extent
★★★★★	20–40	Featured unique ratchet mechanism starter
★★★★★	12–25	First Case tractor to employ roller chain and sprocket drive
★★★★	10–20	Three-wheeler; featured first four-cylinder Case motor

Although the first gasoline-powered tractors appeared as early as 1889, little consensus or standardization of design emerged within the industry before 1920.

Manufacturers of the 1890–1920 period used a multiplicity of designs and components that included the following: one-, two-, three-, and four-cylinder vertical engines;

The massive Model 30–60, introduced in 1912, weighed more than 12 tons.

one- and two-cylinder horizontal engines; two- and four-cycle engines; transverse- and parallel-mounted engines; one-, two-, and three-forward-speed transmissions; one-, two-, and four-wheel-drive systems; open and closed chain- and gear-driven final drives; live and fixed rear axles; and a variety of carburetion, lubrication, ignition, and air filtration systems.

The four earliest Case gasoline tractors are evidence of this confusion. The Model 30–60 and Model 20–40 closely resembled the Case steam traction line—in effect, steam tractor chassis and drives fitted with internal-combustion, two-cylinder gasoline engines. Yet, the two engines were of different design.

The Model 12–25 was smaller and somewhat streamlined, a response to the demand for lightweight tractors. It too featured a two-cylinder engine, whereas the Model 10–20 was a three-wheeler with an automotive type, four-cylinder engine.

The Model 30–60 two-cylinder horizontal engine. Its flywheel alone weighed 1,000lb.

All four models appeared between 1912 and 1916. The fact that these models were dissimilar and perhaps even out-of-date by 1917 does not mean that Case engineers were incompetent. The differences merely reflect the general chaos in the industry and the speed at which the gasoline tractor evolved.

A Model 30–60 photographed September 1992 at the 150th Case anniversary celebration, part of the 39th annual Western Minnesota Steam Thresher's Reunion (WMSTR) held in Rollag, Minnesota.

Model 30–60

J. I. Case Company's first production gasoline tractor, the 30–60, was a "heavy-weight" tractor, which in many ways resembled Case steam traction units. Although recognized for excellence in design and construction at the 1912 Winnipeg Motor Contest, the 30–60 was never as popular as comparable Hart-Parr or Rumley models.

The 30–60 was powered by a transverse-mounted, horizontal, two-cylinder engine that operated at 365rpm (early units at 350rpm). It can only be described as massive. Its cylinder bore measured 10in, with a stroke of 12in. The crankshaft journals measured $4\frac{1}{2}$in in diameter; the six-spoked flywheel weighed approximately 1,000lb; and the belt pulley measured $32\frac{1}{2}$in in diameter with a $12\frac{1}{2}$in face. Case rated the engine at 60 belt hp and promoted it at a maximum rating of 75hp. At the drawbar, output measured 30hp.

Like most tractors of its day, the 30–60 could be operated on gasoline or a variety of low-grade fuels including naphtha, kerosene, and other distillates. The tractor was started on gasoline, after which the operator switched over to the low-grade fuel. To accommodate both fuels, two tanks were placed beneath the 30–60 operator platform, the larger for distillates and the smaller for gasoline.

The 30–60 employed a low-tension or low-voltage make-and-break ignition system, with an oscillating Sumter magneto. The tractor drew electricity from batteries for starting, after which the operator switched over to the magneto.

Cooling was achieved by induced draft. Heated exhaust gasses were piped away from the engine to the top of the cooling tank mounted at the front of the chassis. Meanwhile, a centrifugal pump circulated water through the engine and the tubular radiator fitted inside the cooling tank. A draft was created above the radiator, as the hot exhaust gasses escaped from the top of the cooling

The Model 30–60 belt-driven water pump. The 30–60 cooling system capacity was 170gal.

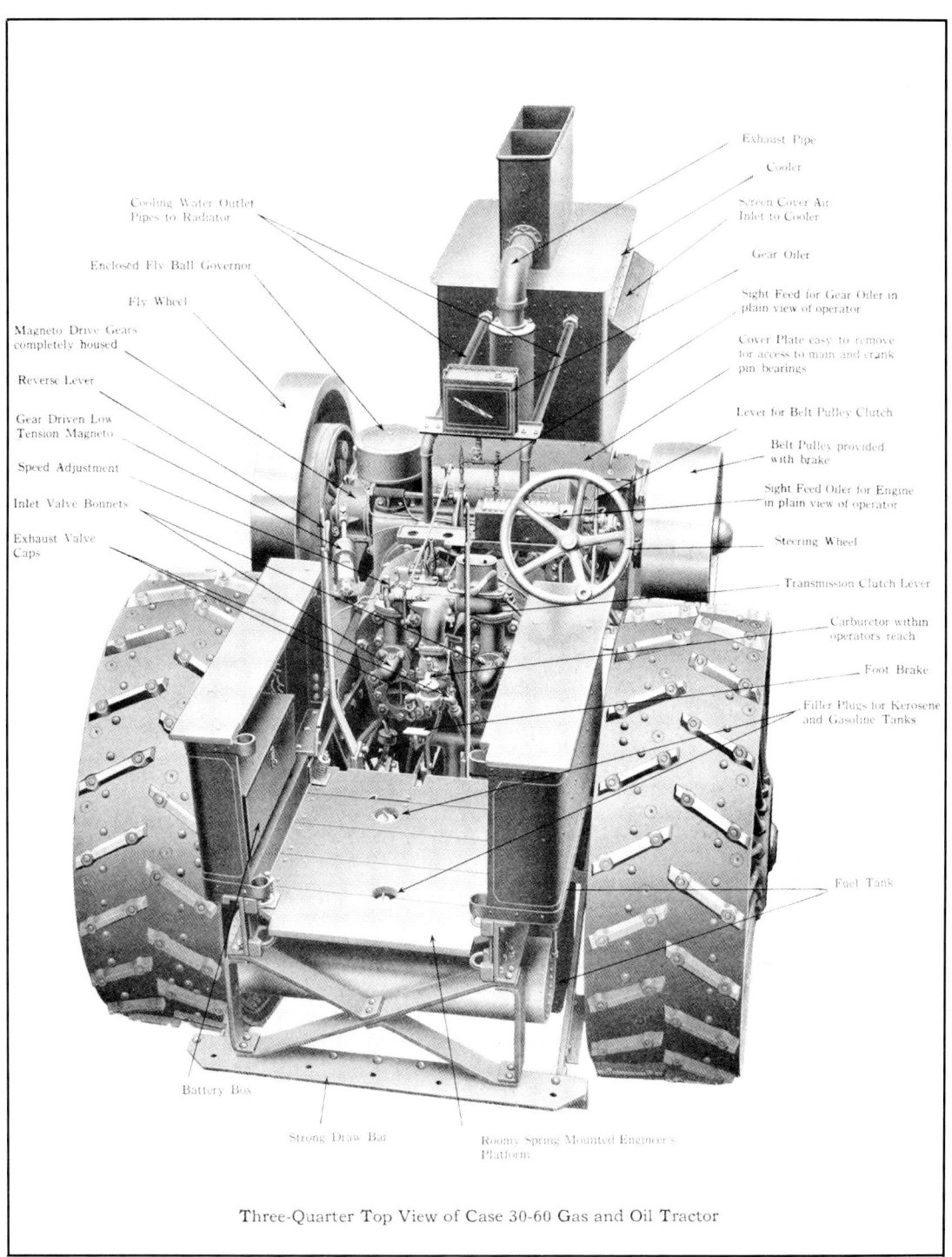

Cooling Water Outlet
Pipes to Radiator

Enclosed Fly Ball Governor

Fly Wheel

Magneto Drive Gears
completely housed

Reverse Lever

Gear Driven Low
Tension Magneto

Speed Adjustment

Inlet Valve Bonnets

Exhaust Valve
Caps

Exhaust Pipe

Cooler

Screen Cover Air
Inlet to Cooler

Gear Oiler

Sight Feed for Gear Oiler in
plain view of operator

Cover Plate easy to remove
for access to main and crank
pin bearings

Lever for Belt Pulley Clutch

Belt Pulley provided
with brake

Sight Feed Oiler for Engine
in plain view of operator

Steering Wheel

Transmission Clutch Lever

Carburetor within
operators reach

Foot Brake

Filler Plugs for Kerosene
and Gasoline Tanks

Fuel Tank

Battery Box

Strong Draw Bar

Roomy Spring Mounted Engineer's
Platform

Three-Quarter Top View of Case 30-60 Gas and Oil Tractor

This three-quarter overhead view gives a definite impression of the size and mass of the Model 30–60. The rear wheels illustrated were the standard "traction" type, the same as fitted to Case steam tractors.

15

Model 30–60 Specifications

Recommended for six or eight 14in plows; 36in cylinder thresher fully equipped

Belt Horsepower: 60
Drawbar Horsepower: 30–35
Cylinders: Two: bore, 10in; stroke, 12in
Lubrication: Force feed by sight feed pump
Cooling System: Centrifugal pump circulating system; capacity, 170gal
Belt Pulley: Speed, 365rpm (early units 350rpm); diameter, 32in; face, 12½in
Front Wheels: Diameter, 42in; width, 12in
Rear Wheels: Diameter, 72in; width, 24in; extension rims, 12in
Wheelbase: 126in
Overall Width: 105in
Overall Height: 126in
Weight: 25,800lb
Drawbar Height: 25¾in
Platform Height: 50in
Height to Center of Crankshaft: 60in
Overall Length: 223in
Road Speed: 2mph
Fuel Capacity: Gasoline, 22gal; kerosene, 90gal

tank. The draft pulled outside air across the radiator tubes and cooled the water.

The 30–60 featured a heavy chassis, constructed from hot-riveted, 10in channel steel. As with most early tractors, the 30–60 required a massive backbone to support its large, heavy components.

The 30–60 expanding-shoe clutch was mounted inside the flywheel. Power was transferred to the first transmission gear from a gear on the crankshaft positioned behind the clutch. The transmission featured one forward speed of 2mph and a reverse speed.

Model 30–60 Production Totals by Model Year
Source: J. I. Case Company

1912	1913	1914	1915	1916	Total
125	294	—	36	38	493

A Model 30–60 prototype photographed in 1911.
Case Company

Case supplied its dealers with this advertising
electrotype for local promotion of the Model 30–60.

Front wheels measured 42x12in. The standard rear wheels measured 72x24in. Optional 12in extension rims were available.

Most experts agree that the 30–60 was not actually manufactured by Case Company. Rather, it was built under contract by Minneapolis Steel and Machinery Company, manufacturer of the Twin City line of tractors.

Introduction of the Model 30–60 is sometimes dated at 1911. However, Case Company records indicate full-scale production began in 1912 and continued into 1916. Sales of large tractors fell off sharply after 1913, as demand moved away from the early design of mammoth heavyweights toward the smaller, more agile tractors referred to as "lightweights."

The 30–60 was priced at $2,500 in 1912 and at $2,800 by 1914. Inventories of the 30–60 were carried over beyond 1916. The tractor was featured in advertisements and Case catalogs as late as 1918.

Collecting Comments

Early tractors produced in limited quantities generally command high prices. The 30–60 is no exception. As with any tractor of this era, it is also expensive to restore. Assume that replacement parts are impossible to find and, therefore, must be custom cast or machined. A large, well-equipped shop is required to handle its massive components, and the tractor is too heavy to be easily transported. For such practical reasons, the 30–60 is not attractive to the average collector.

Yet, to the enthusiast who can afford such tractors, the 30–60 is a valuable machine that will only become more so, as interest in collecting Case tractors grows.

The Model 30–60 is a tease, in the sense that as few as five units are known to exist. If,

This well-maintained Model 30–60 is owned by the Western Minnesota Steam Thresher's Reunion.

however unlikely, you do uncover a forgotten 30–60, don't tell a soul until you secure a bill of sale from its present owner. Then, give me a call.

Model 20–40

The Model 20–40 was introduced in 1912 and was built through 1919. Although it weighed close to 14,000lb, it offered an uncomplicated design and relative fuel economy. It gained an excellent reputation among farmers who required a five- to six-plow tractor.

The 20–40 was introduced with a two-cylinder, horizontal, opposed valve-in-head engine of 7¾in bore and 8in stroke. The engine was mounted transversely. Some experts contend that this early engine was manufactured by Davis Motor Works of Milwaukee, Wisconsin. In 1913, a larger engine (Case-built) of 8in bore and 9in stroke was introduced. In 1916 or 1917, the bore was

The Model 20–40 two-cylinder opposed engine. Bore and stroke of the 1912–1913 engine measured 7¾inx9in; 1913–1916 engine, 8inx9in; and 1916–1919 engine, 8¾inx9in.

increased to 8¾in. In spite of these changes, the engine was consistently rated by Case at 40 belt hp and 20 drawbar hp.

The engine flywheel was positioned inside the chassis member and behind the belt

Catalog illustration of a 1917 Model 20–40. The cab with the canopy over the engine, the enshrouded truck-type radiator, and the carburetor placed below the frame side member identify this as a later model.

Model 20–40 Specifications

Recommended for six 14in plows, 32inx54in thresher with all attachments

Horsepower: 40
Drawbar Horsepower: 20–25
Cylinders: Two: bore, 8³/₄in; stroke, 9in
Lubrication: Madison-Kipp force sight feed
Cooling System: Capacity, 24gal
Belt Pulley: Speed, 475rpm; diameter, 24in; face, 8½in
Front Wheels: Diameter, 40in; width, 10in
Rear Wheels: Diameter, 66in; width, 20in; extension rims, 8 or 12in
Wheelbase: 114in
Overall Width: 100in
Overall Height: 108½in
Weight: 13,700lb
Drawbar Height: 24½in
Platform Height: 30³/₈in
Height to Center of Crankshaft: 55in
Overall Length: 177in
Road Speed: 2mph and 3mph
Fuel Capacity: Gasoline, 11gal; kerosene, 26gal

pulley, on the right-hand side of the tractor. The tractor was started through the use of a ratchet mechanism, unique to Case, that was fitted to the end of the crankshaft, opposite the flywheel.

The 20–40 employed a Kingston carburetor. Engine lubrication was force fed by a pump, driven by an eccentric on the camshaft. The Madison-Kipp system featured a multiple feed oiler located in front of the operator. Small windows in the top of the oiler permitted a view of its operation.

Model 20–40 Production Totals by Model Year

Source: J. I. Case Company

1912	1913	1914	1915	1916	1917	1918	1919	Total
796	1173	227	165	500	602	500	300	4263

Valve mechanism on 1912 Model 20–40. This engine may have been built by Davis Motor Works of Milwaukee, Wisconsin.

The fully enclosed, fly ball governor fitted to the engine was designed and built by Case. It regulated engine speed at 450rpm on the Davis engine and at 475rpm on later units. The 20–40 employed a high tension or high voltage "jump-spark" ignition system that featured a K-W magneto with impulse starter. It used $^3/_4$in pipe size plugs.

The early 20–40 was fitted with a tubular-type radiator. As with the 30–60, exhaust-induced draft pulled cool air across the radiator tubes. The system did not include a water pump. Rather, it was dependent on thermosyphoning to circulate the water.

In 1916, the cooling system was modified. A cellular or truck-type radiator with large lower and upper water tanks replaced the tubular-type radiator. Draft for the radiator was augmented by a fan, operated by a friction wheel driven by the flywheel. Water circulation remained dependent on thermosyphoning.

The 20–40 chassis was built of hot-riveted, 8in steel channel sections. Like that of the 30–60, it was designed to carry the weight of its heavy components.

Shifting gears was accomplished by means of a hand clutch lever and single shift lever. The expanding-shoe friction clutch served both for traction and belt pulley operations. Road speeds in two forward gears were 2mph, for plowing and heavy work, and $2^3/_4$mph (3mph on later units), for hauling and other road work. Reverse speed was 2mph.

In the early days of the farm tractor there was no sanctioned testing. However, beginning in 1908 and carried out through 1913, the Winnipeg Industrial Exhibition sponsored an annual Agricultural Motor Competition. Score sheets were published and favorable results from these trials were widely used by manufacturers to promote their tractors. In 1912, the 20–40 won a gold medal in its class.

The Model 20–40 featured a ratchet starting mechanism unique to Case.

As could be expected, this fact was prominently featured in Case advertising.

In 1919, in an effort to protect Nebraska farmers from poorly built tractors, the Nebraska legislature enacted a law requiring all tractors sold in the state to be "tested and passed upon by a board of three engineers in the employ of the state university (at Lincoln)."

Instituted in 1920, the Nebraska Tractor Tests provided farmers with an objective measure of comparison. Their impact was significant. The tests accelerated tractor improvements and expedited the death of a number of inferior tractors.

Although production of the 20–40 ended in 1919, Case Company submitted the tractor for testing at Nebraska in 1920 (Official Tractor Test No. 07). The test tractor weighed in at 13,780lb. It generated slightly over 40 belt hp under rated load and recorded drawbar pull of 3,987lb. Under maximum load, the tractor transferred 24.66hp to the drawbar and recorded drawbar pull of 5,537lb.

The 20–40 underwent several changes through the course of its production. Some changes were obvious, making it easy to distinguish between earlier and later tractors.

The earliest units carried the operator's platform and cab set on top of the channel iron frame. The cab itself featured a heavy pillar positioned in the middle of the open frame, at the front of the cab. In later units, the platform and cab were set between the frame side members and therefore sat lower on the frame. The wide post in the cab window was eliminated, and the roof line was extended forward to completely cover the engine.

The 20–40 carburetor was changed from a unit mounted above the frame side member on earlier units to one mounted below the side member. The lower placement improved the feed of fuel from the tanks.

The exposed, tubular-type radiator would indicate this Model 20–40 was built prior to 1916. *Shields Library, Special Collections, University of California, Davis*

This 1918 factory photograph shows a Model 20–40 with truck-type radiator. *Case Company*

The flywheel, clutch mechanism, and belt pulley of a Model 20–40. The belt pulley measured 8$\frac{1}{2}$x24in.

The Model 20—40, like the Model 30—60, featured a heavy channel-steel frame and transmission similar to that of a Case steam tractor.

Clutch Side of Motor

The expanding-shoe clutch mechanism of the Model 20—40 was positioned inside the belt pulley.

As mentioned, the radiator fitted to early units was a tubular type that was fully exposed. It was soon enshrouded and later replaced by a truck-type radiator. Again, units with the earlier style radiator had no cooling fan, while fans were added to improve cooling with the later style radiator.

Yes, the 20—40 was big. However, it was an extremely durable and efficient tractor with impressive drawbar pull. It proved a popular tractor, as evidenced by the number of units built.

The 20—40 retailed for $2,000 in 1915. The tractor was included in the 1920 General

The early Model 20—40 featured a cab set on top of the frame. Note the road bands on the rear wheels.
Case Company

Exhaust Piping

Flexible Rubber Hose Connection

Large Water Piping for Thermo Syphon Cooling

Magneto easy to get at from ground

Sensitive Fly Ball Governor

Multiple Feed Oiler in plain view of operator

Safety Starting Device

Cylinders separate, bolted to crank case

Steering Wheel

Engine Crank Case solid bolted to frame channels

Fuel Tank Filler Plug

Fuel Tank Gravity Feed to Carburetor

Top Cover Plate easy to remove without disturbing cam shaft or magneto setting to gain access to inside of crank case

Only one Lever for two road speed and reverse

Out Board Bearing to take stress of belt pull

Brake on Belt Pulley

One Clutch Lever for traction or belt work

Easy Adjustment of Clutch shoes

Brake directly on differential

Oil Cup for Countershaft Bearing Lubrication

Strong Drive Wheels set close to bearings, no overhang on rear axle

Tubular Counter Bearing for rear axle. Note bearing to cross member.

Swinging Draw Bar pivoted under rear axle

Roomy Platform for engineer

Three-quarter Top View Case 20-40 Gas and Oil Tractor

Illustration of a 1917 Model 20–40. Note the later-style radiator and cooling fan.

25

Most 20–40s featured the cab positioned between frame side members, such as this tractor photographed at the 1992 WMSTR event.

The earliest Model 20–40 featured the carburetor mounted above the frame.

Catalogue at $2,500, but was absent from the 1921 Farm Tractor catalog.

Collecting Comments

While there were significantly more units of the 20–40 built than the 30–60, it is still a rare and valuable tractor. Like the 30–60, it was a large machine with heavy components that are difficult to handle and nearly impossible to replace. Restoration costs are high.

You are unlikely to stumble across a restored 20–40 that is not known to established Case collectors. Yet, a number of 20–40s are still around. If you are determined to own a heavyweight gas tractor, the 20–40 is an excellent choice.

Relatively speaking, the earliest production units (those fitted with the 7¾in bore engine) are perhaps the most valuable. However, any 20–40 in a restorable condition would be a desirable find.

The carburetor of the Model 20–40 was soon moved below the frame and positioned lower than the fuel tanks.

A 1912 Model 20–40 photographed at the 1992 WMSTR event.

Model 12–25

The Model 12–25, introduced in 1914 and built through 1918, was targeted at farmers seeking a smaller Case tractor. Its open operator platform, low-profile, square radiator, and fully enclosed engine compartment presented a streamlined look that set it apart from the larger 30–60 and 20–40.

The market for the 12–25 may have been limited by competition from the Case 20–40.

Case ready-made ad 3-726 for the Model 20–40.

A Model 20–40 photographed at the 1992 WMSTR
event. Note the optional rear wheel extensions.

An illustration of the Model 12–25 from the 1917
catalog.

Also, at 8,995lb, the 12–25 was too large a tractor to compete with the likes of the popular International Harvester Titan 10–20, Wallis Cub, and the growing number of light-weight tractors that emerged in this period.

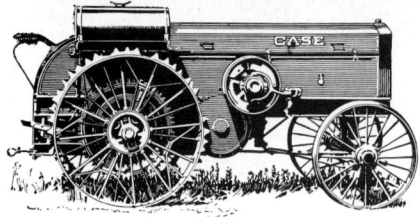

Investigate Case Tractors First

WE handle the Case Kerosene Tractors. Above is shown the 12-25. We can also supply you with larger sizes —the Case 20-40 or the 30-60.

This 12-25 Case is for all-round utility work on almost any farm. It pulls a four-bottom plow, drives a Case 26x46 thresher completely equipped, pulls two 7 or 8 ft. binders, runs a No. 16 Case Silo or a Case 8x15 Rock Crusher. In fact it does any kind of work requiring this amount of power.

We recommend this tractor with full confidence. You cannot find a better one. It has been a success all over the country. Thousands are in use today.

Before you buy a tractor, you will do well to investigate the Case line. You know the J. I. Case Threshing Machine Company. It has been in business for 76 years, so you can rely on their engineers to give you the latest and best. And you'll not be experimenting.

Come in and let's talk tractors.

Your Name Here
Your Address

The Sign of
Mechanical
Excellence
the World
Over

Case ready-made ad 3-724 for the Model 12–25.

The 12–25 featured a two-cylinder, horizontal, opposed valve-in-head motor with 6¾in bore and 7in stroke. In 1915, the bore was increased to 7in. The Case-built governor, driven by a spur gear off the camshaft, regulated engine speed at 600rpm. As with all Case tractors built prior to 1929, the engine was mounted transversely. Case rated the engine at 25 belt hp and between 12 and 15 drawbar hp.

The engine was fitted with a Kingston carburetor and Kingston high-tension magneto with impulse starter. Engine lubrication was force fed by a Madison-Kipp sight feed system like that used on the larger Case tractors. Cooling was by thermo syphoning and aided by a fan.

The 12–25 frame was built from steel channels and diagonally braced behind the front axle. The hitch was placed lower than the rear axle, with the drawbar positioned about 18in above the ground.

Model 12–25 Production Totals by Model Year
Source: J. I. Case Company

1913	1914	1915	1916	1917	1918	Total
2	100	503	933	721	1062	3321

Model 12–25 Specifications
Recommended for four 14in plows, 24inx46in thresher with all attachments

Horsepower: 25
Drawbar Horsepower: 12–15
Cylinders: Two: bore, 7in; stroke, 7in
Lubrication: Madison-Kipp force sight feed
Cooling System: Capacity, 15gal
Belt Pulley: Speed, 600rpm; diameter, 22in; face, 7½in
Front Wheels: Diameter, 38in; width, 8in
Rear Wheels: Diameter, 56in; width, 18in; extension rims, 6 or 8in
Wheelbase: 90in
Overall Width: 73in
Overall Height: 76in
Weight: 8,995lb
Drawbar Height: 18in
Platform Height: 24in
Height to Center of Crankshaft: 44½in
Overall Length: 149in
Road Speed: 1¾mph and 2⅛mph
Fuel Capacity: Gasoline, 17gal; kerosene, 17gal

One of the first two Model 12–25 tractors built.
Case Company

The later Model 12–25 featured fuel tanks mounted above the rear wheels. *Shields Library, Special Collections, University of California, Davis*

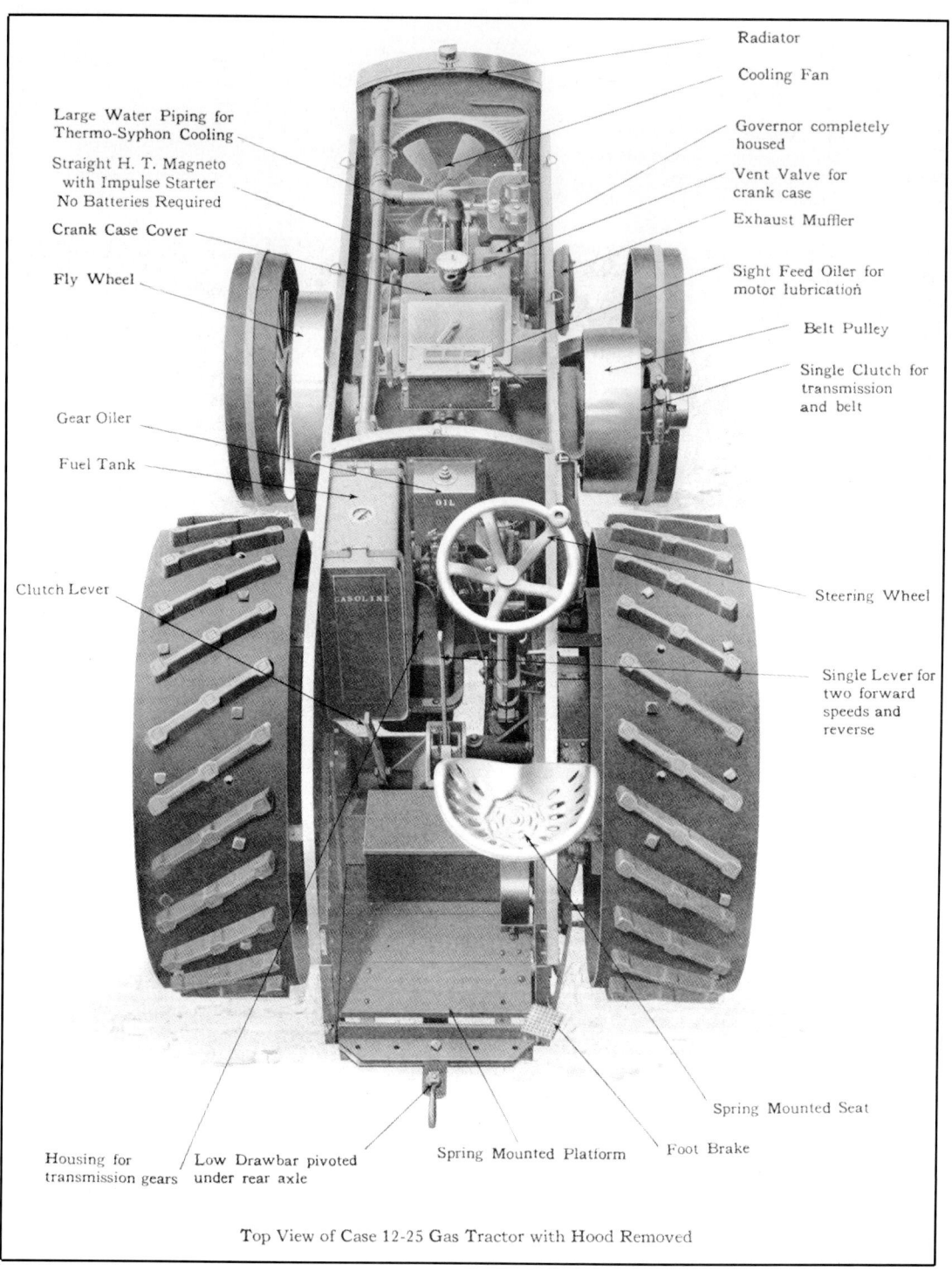

Radiator

Cooling Fan

Governor completely housed

Vent Valve for crank case

Exhaust Muffler

Sight Feed Oiler for motor lubrication

Belt Pulley

Single Clutch for transmission and belt

Large Water Piping for Thermo-Syphon Cooling

Straight H. T. Magneto with Impulse Starter No Batteries Required

Crank Case Cover

Fly Wheel

Gear Oiler

Fuel Tank

Clutch Lever

Steering Wheel

Single Lever for two forward speeds and reverse

Spring Mounted Seat

Housing for transmission gears

Low Drawbar pivoted under rear axle

Spring Mounted Platform

Foot Brake

Top View of Case 12-25 Gas Tractor with Hood Removed

The Model 12–25 was relatively compact in comparison to its stablemates. Still, the tractor weighed 9,000lb.

Unlike other tractors in the Case line, which employed gears to transfer power from the crankshaft to the transmission, the 12–25 employed sprockets and nickel-steel roller chain. The chain and sprockets were entirely enclosed by a sheet metal casing and ran constantly in oil.

The differential gears, reverse gears, and bull pinion gears were housed in a cast-iron case, and also ran constantly in oil. The transmission featured two forward speeds of 1^3/$_4$mph and 2^1/$_5$mph. The clutch, which served both for traction and for operation of the belt pulley, was an expanding-shoe friction-type engaged by a hand lever.

The 12–25 was priced at $1,300 in 1916. (By way of comparison, the Waterloo Boy Model N, also rated at 25 belt and 12 drawbar horsepower, sold for $1,150 in 1917.) Inventory carried over to 1921 was priced at $1,425.

Collecting Comments

The 12–25 is representative of the tractors built when manufacturers first sought to

The Model 12–25 channel-steel frame.

Rear view of an early Model 12–25. The fuel tank was positioned on the left side of the operator platform on early units.

Rear view of a later Model 12–25.

The Model 12–25 transmission featured two forward speeds.

broaden their market by scaling down the size of their machinery.

Best described as a "light" heavyweight, the 12–25 is my favorite among this early series of Case tractors. It is a handsome machine with a simple line but a strong appearance. It was a more manageable tractor in the field than many of its contemporaries, which makes it well-suited to the collector who seeks an early gas tractor for parades and public display.

As with any tractor, be wary of an engine that is not running or appears to be weak. These engines were built in limited numbers, parts and restoration costs are high.

Model 10–20

Introduced in 1915, the Model 10–20 was built through 1918. Priced under $900, it was targeted for sale to the average farmer. Its three-wheel design was unconventional but hardly original. The 10–20 closely resembled the Bull Tractor Company's Little Bull, a tractor that dominated the market in 1914.

An early Model 12–25 photographed at Rollag.

The engine side panels of the Model 12–25 gave
this tractor a more streamlined look.

A second Model 12–25 photographed at the 1992
WMSTR event.

An illustration of the three-wheel Model 10–20 from
the 1917 catalog.

The 10–20 featured the first vertical four-cylinder, valve-in-head Case engine. With cylinders cast en bloc, this modern, light-weight motor developed 20 belt hp at 900rpm. Cylinder bore measured $4^{1}/_{4}$in; stroke measured 6in. In its Nebraska Tractor Test of 1920 (Official Tractor Test No. 06), the 10–20 demonstrated maximum drawbar pull of 2,631lb at 15.28 drawbar hp.

The motor was fitted with Kingston carburetor and Kingston high-tension magneto with impulse starter. Lubrication was by a plunger pump driven by the camshaft. It fed oil to the main bearings and to splash shields underneath each crank pin. Cooling was by means of a centrifugal pump and fan.

The 10–20 featured a single wheel in front and two rear wheels of like diameter but different widths. Power was transferred through three sets of spur gears to the wider, right rear wheel only. As only one wheel was driven, the one-speed transmission did not require a differential. However, a special "Jaw Clutch" was fitted to the left rear wheel and

axle that, when engaged, transferred power to both wheels. Case promoted this feature as "very important . . . in a tractor of this type, as in hard places it gives additional traction, and should the idler wheel drop in a hole or rut, it would be impossible to get out without this feature."

The simple drivetrain offered the primary advantage of being less expensive to manufacture. While the 10–20 was the lowest priced Case model, it still sold for more than twice the price of the Little Bull.

It is unlikely that Case would have developed a three-wheeler had it not been for the success of the Little Bull. Promoted as the "Bull with the Pull," the Little Bull was underpowered, suffered other design weaknesses, and faced production problems that

Model 10–20 Production Totals by Model Year

Source: J. I. Case Company

1915	1916	1917	1918	Total
203	1551	2050	2875	6679

The Model 10–20 featured the first Case four-cylinder, valve-in-head engine.

Top View showing Frame Construction

A catalog illustration of the three-wheel frame and driveline of the Model 10–20.

A Model 10–20 and two Case binders in operation.
Case Company

Three-wheelers with power to one wheel were popular for only a brief period, beginning in 1913. Case carried over inventories of the Model 10–20 for three years beyond the end of its production.
Case Company

eventually led to its demise. However, thousands of units were sold—more than 4,000 in 1914 alone. Within a year of its introduction, the Little Bull had captured 40 percent of the tractor market.

It is not surprising that Case would have been tempted by such success. Yet, interest in three-wheelers quickly faded. By 1918, the Fordson ushered in a new era of small tractor design. Case Company, as well, had developed a new series of smaller tractors that would carry the company through the decade of the 1920s.

Judging by production figures, the 10–20 was not a total failure. However, the figures are deceptive. Inventories of the 10–20 were carried over into 1921.

The 10–20 was priced at $890 in 1916. Units carried over to 1921 were priced at $900.

Model 10–20 Specifications
Recommended for three 14in plows, 20inx36in thresher with all attachments

Horsepower: 20
Drawbar Horsepower: 10–12
Cylinders: Four: bore, 4¼in; stroke, 6in
Lubrication: Pressure feed and splash
Cooling System: Capacity, 11gal
Belt Pulley: Speed, 900rpm; diameter, 17in; face, 6½in
Front Wheels: Diameter, 30in; width, 8in
Rear Wheels: Diameter, 52in; width, 22in (drive wheel), 10in (idler wheel)
Wheelbase: 76in
Overall Width: 67in
Overall Height: 60in
Weight: 5,080lb
Drawbar Height: 11in
Platform Height: 21¼in
Height to Center of Crankshaft: 27½in
Overall Length: 158in
Road Speed: 2¼mph
Fuel Capacity: Gasoline, 2¾gal; kerosene, 20gal

Collecting Comments

Three-wheeled tractors played a brief but key role in the history of the tractor. In addition to Bull Tractor Company and Case, Massey-Harris (whose version was built by Bull), Allis-Chalmers, Gehl Bros., and Chase were among the companies who marketed similar tractors.

Among Case tractors, the 10–20 is relatively rare. Yet, it does not seem to attract the same amount of interest as its contemporaries. Nonetheless, I consider it to have excellent investment potential.

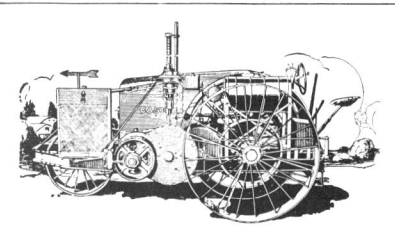

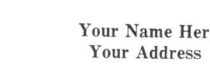
This 1918 ad for the Model 10–20 stressed its popularity among farmers worldwide. An unknown number of units were exported.

Chapter 4

Crossmotor Tractors 1916–1928

Models 9–18, 9–18B, 10–18, 12–20, 15–27, 18–32, K, 22–40, 25–45, T, 40–72

Rating	Model	Remarks
★★★	9–18	The first Case crossmotor
★★★	9–18B/10–18	First one-piece cast-iron frame
★★★★	12–20/A	Check condition of the exhaust manifold carefully
★★★	15–27/18–32/K	Units with factory options, such as canopy, axle extension, acetylene headlights are the most desirable
★★★★	22–40/25–45/T	Low production means greater value, but restoration costs are high
★★★★★	40–72	Big is beautiful and rare is rare

Although many early gasoline tractors were built to pull six or more plow bottoms, the market for tractors of this size was limited. The vast majority of farmers could neither afford nor did they require a tractor of this size.

By the mid–1910s, manufacturers began to focus their attention on the vast potential

"Case Tractors For Road & Farm Work."

An illustration of the Model 9–18 from the 1917 catalog.

that existed for tractors designed to handle two-, three-, and four-bottom plows. Case was no exception. In late 1916, the company introduced the Model 9–18, a lightweight, streamlined, two-plow tractor.

The 9–18 is considered the first in a series of tractors referred to as the Case "crossmotors" or "crossmounts." While the steam era belonged to Case Company, it was this series of rugged, stocky machines that established Case as a major gasoline tractor manufacturer in the 1920s.

Model 9–18

The Model 9–18 was a conventional four-wheel tractor, introduced in 1916 and built through 1918. It featured a vertical four-

cylinder valve-in-head motor, cast en bloc; with cylinder bore of 3⅞in and stroke of 5in. Its fly ball-type governor regulated engine speed at a rated 900rpm.

Model 9–18 Specifications
Recommended for two 14in plows, 20in cylinder thresher without feeder or wind stacker

Horsepower: 18
Drawbar Horsepower: 9–10
Cylinders: Four: bore, 3⅞in; stroke, 5in
Lubrication: Force feed and splash
Cooling System: Capacity, 9gal
Belt Pulley: Speed, 900rpm; diameter, 14¼in; face, 5¼in
Front Wheels: Diameter, 30in; width, 6in
Rear Wheels: Diameter, 48in; width, 10in; extension rims, 6in
Wheelbase: 72in
Overall Width: 58in
Overall Height: 61in
Weight: 3,770lb
Drawbar Height: 15½in
Platform Height: 22in
Height to Center of Crankshaft: 26¾in
Overall Length: 123in
Road Speed: 2¼mph and 3½mph
Fuel Capacity: Gasoline, 2¾gal; kerosene, 16gal

Model 9–18 and 10–18 Production Totals by Model Year
Source: J. I. Case Company

Model	1916	1917	1918	1919	1920	Total
9–18	22	2980	2800	—	—	5802
10–18	—	50	2501	4500	2000	9051

Another illustration from the 1917 catalog showed units fitted with optional Case Air Washer and exhaust muffler.

The Model 9–18 featured a Kingston magneto with impulse coupler.

The motor was fitted with Kingston gasoline or kerosene carburetor and Kingston high-tension magneto with impulse starter. Lubrication was by plunger pump to the main bearings, and by overflow and splash trays to the cylinders and other working parts. The cooling system relied on a centrifugal pump and fan. Later units also featured a Case-designed water air filtration system (see Model 10–18 for details).

The engine was mounted transversely (hereafter the term is "crossmounted") on its frame, built up from 5in steel channel. Side panels and a sloping hood gave the 9–18 a modern, styled look. The crossmounted engine permitted Case to design a fully enclosed, simple, efficient two-speed transmission, composed entirely of straight-spur gears. Forward speeds were 2$\frac{1}{4}$mph and 3$\frac{1}{2}$mph. Gear selection was by sliding-spur gear. The clutch was an expanding shoe-type, activated by a hand lever.

Left side, or forward-facing, view of the Model 9–18 motor. The exhaust is at the top. The opening directly left of the water pump reveals the spiral gear that drove the cooling fan.

A Model 9–18 plowing, photographed in February 1918. *Case Company*

Side view of a Model 9–18 photographed at the
1992 WMSTR event.

This Model 9–18 featured the Case Air Washer.

Case rated the 9–18 at 18 belt hp and 9 to 10 drawbar hp with drawbar pull of 1,500lb. Case claimed a maximum drawbar pull of 2,100lb at 12hp.

Model 9–18B and Model 10–18

In 1918, Case introduced the Model 9–18B. It was built on a one-piece cast-iron frame, rather than the structural steel frame used on the 9–18. The cast-iron frame permitted a lighter tractor, with no loss in strength. It integrated the front-axle support, engine crankcase, transmission, and rear-axle housings in a design unique to Case.

The Model 9–18B was built for only a brief time. Case boosted its horsepower output by increasing engine rpm and redesignated it as the Model 10–18. The 9–18B and 10–18 shared the same cast-iron frame and components. A review of the 10–18 is presumed to cover the significant features of the 9–18B.

Although its displacement was equal to that of the 9–18 engine, the 10–18 engine differed in a number of ways. The lower half of the crankcase was cast as part of the unit

Model 10–18 Specifications

Recommended for two 14in plows, an 8ft binder, a 22-shoe grain drill, an 8ft double section disc harrow, or a manure spreader. At road work it will handle a Case No. 3 road grader

Rated Belt Horsepower: 18
Maximum Belt Horsepower: 23–24
Rated Drawbar Horsepower: 10
Maximum Drawbar Horsepower: 12 to 14
Cylinders: Four: bore, $3^{7}/_{8}$in; stroke, 5in
Cooling System: Capacity, 9gal
Belt Pulley: Speed, 1050rpm; diameter, $14^{1}/_{4}$in; face, $5^{1}/_{4}$in
Front Wheels: Diameter, 30in; width, 6in
Rear Wheels: Diameter, 42in; width, 9in; extension rims, 8in and 10in
Wheelbase: 65in
Overall Width: 56in
Overall Height: $54^{1}/_{2}$in
Weight: 3,400lb
Drawbar Height: $13^{1}/_{2}$in
Overall Length: $101^{1}/_{2}$in
Road Speed: $2^{1}/_{4}$mph and $3^{1}/_{2}$mph
Fuel Capacity: Gasoline, 2gal; kerosene, $10^{1}/_{2}$gal

frame and carried the crankshaft. The upper half of the crankcase, with the cylinders cast en bloc, was bolted directly to the frame. A

The Case Archives identified this as a Model 9–18B. *Case Company*

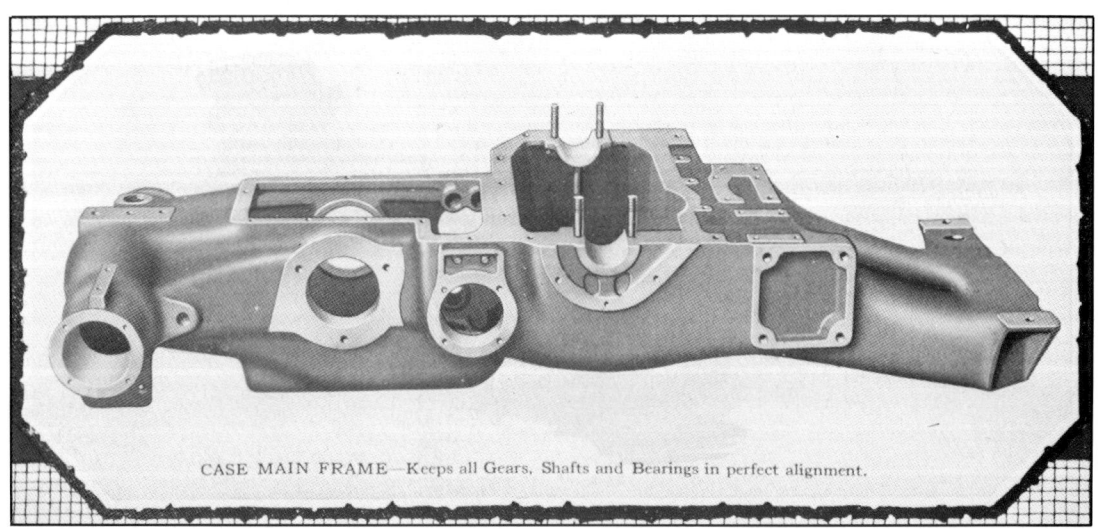

Factory illustration of a Model 10–18. *Shields Library, Special Collections, University of California, Davis*

CASE MAIN FRAME—Keeps all Gears, Shafts and Bearings in perfect alignment.

The Model 9–18B and Model 10–18 cast-iron main frame.

removable pan was bolted underneath the frame.

The crankshaft of the 10–18 was heavier than that of the 9–18 ($2\frac{3}{8}$in diameter versus $2\frac{1}{8}$in diameter). Engine speed was 1050rpm, an increase from 900rpm.

Oil lubrication was originally by pump to the main bearings and by overflow and splash trays to the cylinders and other working parts, just as with the 9–18 and 10–20. However, after serial number 40134, the system was changed to a more modern, full-pressure feed-type with a drilled crankshaft.

The 10–18 also introduced new features in cooling and air filtration. The tractor was fitted with a tube-and-fin-type radiator, in place of a cellular type. The system featured a centrifugal water pump, fan, and for the first time, a thermostat. The thermostat allowed the engine to warm up more quickly as water bypassed the radiator until the temperature reached 160–180 degrees Fahrenheit. According to Case literature, it also kept "the motor hot enough to vaporize kerosene under all loads. . . ."

Case 4-cylinder Valve-in-head Motor

The upper half of the engine, with cylinders cast en bloc, was bolted directly to the frame. The lower half of the crankcase, which carried the crankshaft, was cast as part of the Model 10–18 unit frame.

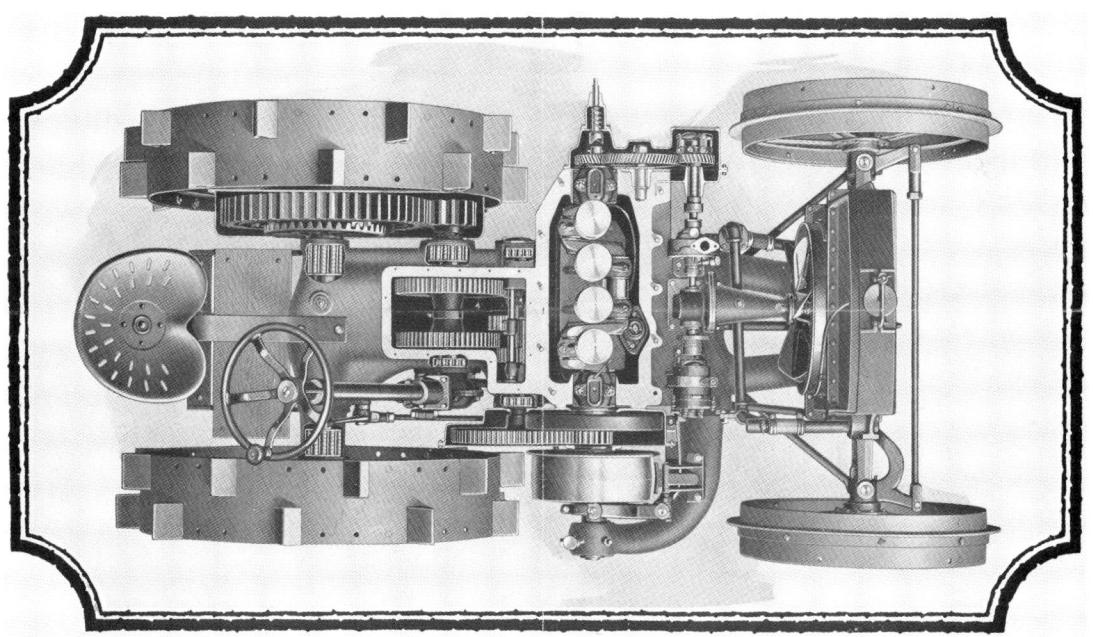

The major components of the Model 10–18 are visible in this cutaway illustration reproduced from the catalog *Farming the Modern Way.*

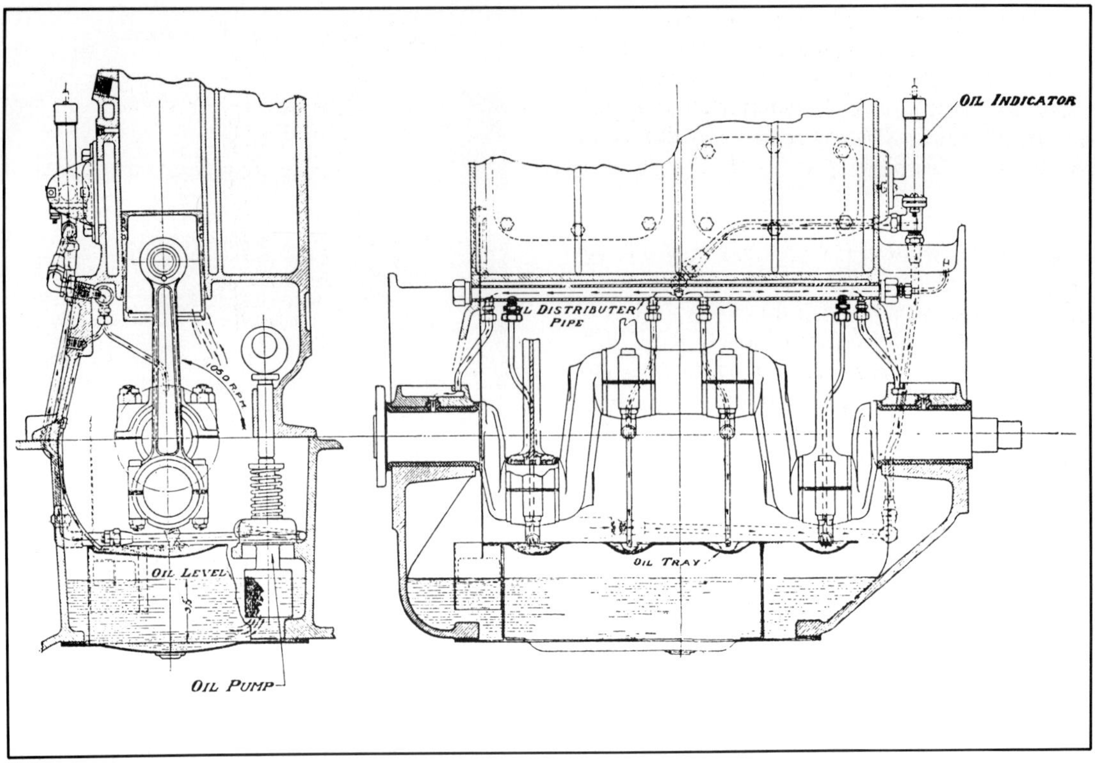

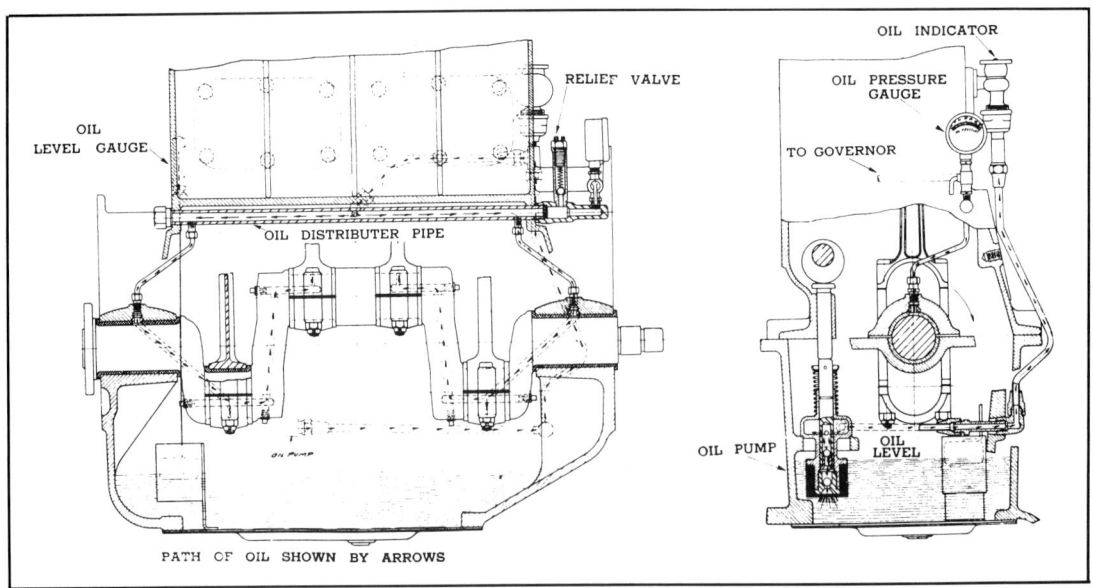

The early Model 10–18 motor featured direct oil feed to main bearings but relied on overflow and splash trays to lubricate the cylinders and other working parts.

The later Model 10–18 engine featured a drilled crankshaft for full pressure feed.

The Case patented Air Washer was also a feature of the 10–18 (offered on later units of the 9–18 as well). Outside air was drawn into a canister and through water that trapped dirt particles. A screen filtered out the dust, as clean air continued through the system into the carburetor.

The 10–18 was tested at Nebraska in April 1920 (Official Tractor Test No. 03). Its engine developed 18.41 belt hp and 11.24 drawbar hp, under maximum load. Maximum drawbar pull was measured at 1,730lb.

The 10–18 was offered in both agricultural and industrial models. Equipped with solid rubber tires, the tractor was sold for use in towing trailers and for light road building and maintenance.

To confuse matters, it would appear that the company built several units of 9–18B after production of the 10–18 began. This overlap of production makes it difficult to determine the exact number of 9–18B units built.

The 10–18 was priced at $1,090 in 1921.

Model 12–20

Introduced in 1922 to replace the Model 10–18, the Model 12–20 was built through 1927. In 1928 it was redesignated the Model A

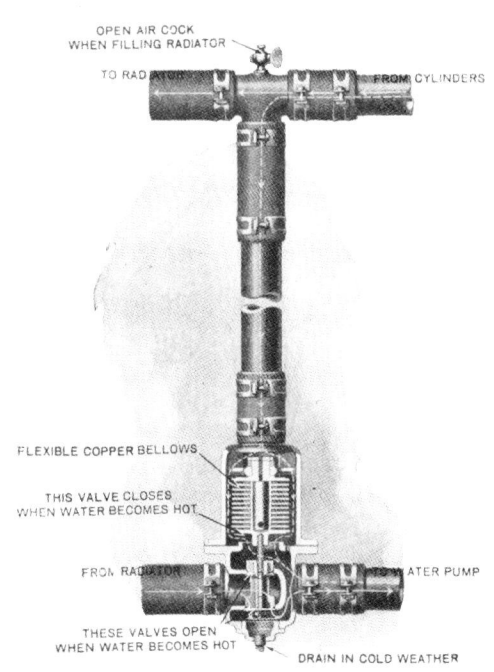

Sylphon Regulator Showing Path of Circulation

The Model 10–18 thermostat kept the engine running at an optimum temperature to vaporize kerosene.

The tube and fin-type radiator introduced to the crossmotor series.

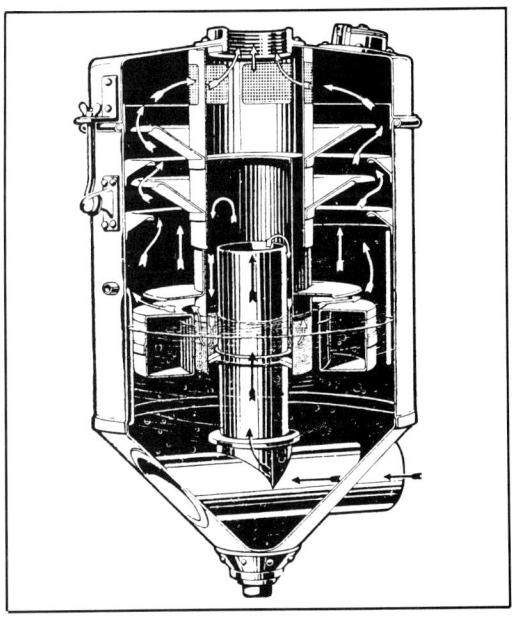

The Case patented Air Washer.

A Model 10–18 and Model 20–28 separator thresh-ing onion seed. *Case Company*

The Model 12–20 featured unique broad-spoked wheels, in this case fitted with standard rear wheel angle-iron grouters. *Shields Library, Special Collections, University of California, Davis*

and remained in production through that year.

While its basic features were similar to those of the 10–18, the 12–20 was not simply a rerated 10–18. From its heavier engine to its broad-spoke wheels, the 12–20 was a different tractor.

A comparison of Nebraska Test results and published specifications point to the differences in these two tractors: the 12–20 outweighed the 10–18 by 690lb; while their wheelbase, overall width, and height were nearly identical, the 12–20's overall length was greater by $7\frac{1}{2}$in; in their respective Nebraska tests, the early 12–20 developed maximum drawbar pull of 2,225lb; the 10–18 developed maximum drawbar pull of 1,730lb.

The 12–20s vertical four-cylinder valve-in-head engine was cast en bloc, and featured removable or renewable cylinder sleeves. Engine bore had been increased to $4\frac{1}{8}$in,

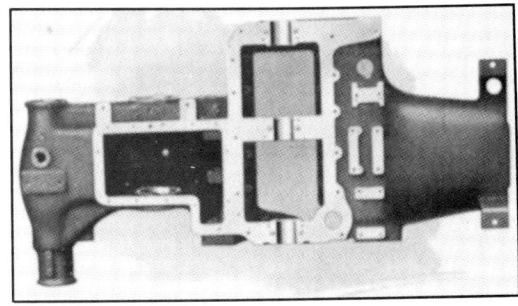

The Model 12–20 one-piece cast-iron frame was lightweight, yet gave the tractor the rigidity it required.

while stroke remained unchanged at 5in. While the crankshaft diameter remained unchanged at $2\frac{3}{8}$in, Case built the 12–20 engine with three main bearings rather than two. Lubrication was pressure fed through the drilled crankshaft.

Model 12–20 with spade lugs and optional 6in extension rims.

At introduction, the engine was fitted with Kingston Model 13-V carburetor and Berling Model F–41 high-tension magneto with impulse coupling. In 1923, Case switched to a Kingston Model L–3 carburetor and Bosch Model At–4 magneto. Both versions of the 12–20 were tested at Nebraska. These changes resulted in improved fuel economy and drawbar performance under maximum load.

Pulley-side view of the Model 12–20.

Cylinder Head and Barrels Are Removable

The Model 12–20 four-cylinder engine was the first Case engine to feature renewable cylinder sleeves.

Model 12–20 Specifications

Recommended for three 14in plows under ordinary conditions, an 8ft double disc harrow, a 22x36in thresher with attachments, a 12ft grain drill and a harrow of equal width

Rated Belt Horsepower: 20
Rated Drawbar Horsepower: 12
Rated Drawbar Pull: 2,045lb
Maximum Belt Horsepower: 25
Maximum Drawbar Horsepower: 14–16
Cylinders: Four; bore, 4¹/₈in; stroke, 5in
Normal Engine Speed: 1050rpm
Ignition: Bosch magneto with impulse coupling; spark plugs, ⁷/₈in SAE standard
Carburetor: Kingston vertical, with single nozzle; size, 1¹/₄in
Fuel Capacity: Gasoline, 2¹/₄gal; kerosene, 17¹/₂gal
Cooling System: Capacity, 10gal
Road Speed: 2¹/₅mph and 3mph; Industrial tractor; 2¹/₄mph and 3mph or 3mph and 4¹/₄mph with optional gears
Overall Length: 109in
Overall Width: 58⁵/₁₆in
Overall Height: 55¹/₂in
Wheelbase: 65in
Shipping Weight: 4,230lb
Drawbar Height: 13¹/₂in
Turning Circle: 24ft
Front Wheels: Diameter, 30in; width, 6in: Industrial tractor fitted with 27x3¹/₂in tire, actual outside diameter 27¹/₂in
Rear Wheels: Diameter, 42in; width, 12in: Industrial tractor fitted with 10x40in tire, actual outside diameter 42in

Model 12–20 Nebraska Test Performance Comparison

Test Measure	Test No. 88 August 1922*	Test No. 91 April 1923**
Belt hp (Rated Load)	20.17	20.16
Fuel Consumption (Rated Load)	2.35 gal/hr	1.996 gal/hr
Drawbar hp (Rated Load)	13.15	12.83
Drawbar Pull (Rated Load)	1,703lb	1,557lb
Belt hp (Maximum Load)	22.51	25.54
Drawbar Pull (Maximum Load)	2,225lb	3,150lb

*12–20 fitted with Kingston 13-V carburetor and Berling magneto
**12–20 fitted with Kingston Type L–3 carburetor and Bosch magneto

The 12–20 featured a unique and quite complex exhaust manifold that was common to several crossmotor models. Designed to improve fuel economy by preheating air in the intake manifold, it featured what Case called "exhaust deflectors." Placed on either end of the manifold, they were controlled by levers. Turning the deflector levers horizontally allowed exhaust gases from all four cylinders to pass around the intake manifold, thereby raising air temperature in the intake manifold. Turning the deflector levers straight up allowed the exhaust from the end cylinders to go through a bypass to the main exhaust without coming in contact with the intake manifold. (It should be noted that the tractor also featured a radiator curtain, which could also be used to regulate engine temperatures.)

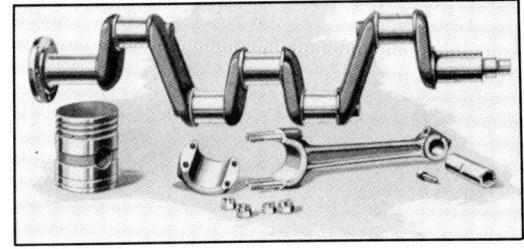

The Model 12–20 engine featured a heavier three-bearing crankshaft, versus the two-bearing crankshaft fitted to the 10–18.

The system also incorporated the Case Air Washer, which featured a damper that directed air toward the manifold for heating or directly to the carburetor.

The amount of water drawn into the carburetor through the water feed valve in-

A-22082-K.T.

Restorer beware! The complex exhaust manifold with heat control was fitted to several crossmotor tractors. In poor condition, it is expensive to repair or replace. *Case Company*

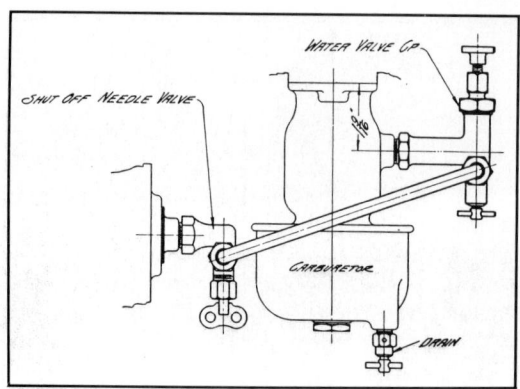

Water feed valves were features of most early tractors. Water was mixed with the fuel to retard ignition.

fluenced performance as well and also required regulation by the operator. Water feed valves were common in this period as a means of mixing water with fuel to retard ignition under heavy work loads or high temperatures.

In the Case system, water was injected through a feed valve with a spray nozzle. Water was drawn directly from the cooling system, through a plug in the cylinder block. The amount of water injected was controlled through an adjustment at the base of the water valve. It was important that the valve be closed when the tractor was started and just before the engine was stopped. With the carburetor choke closed for starting, the

Model 12–20 Production Totals by Model Year
Source: J. I. Case Company

1921	1922	1923	1924	1925	1926	1927	1928	Total
4	300	1325	1475	1800	2500	2000	2450*	11854

*Model A; includes 50 industrial tractors

Directions for setting heat controls for various loads and weather conditions and suggestions for correct operation of engine.					
Note:—This chart while only suggestive, will serve as a guide if operator will use a little care in judging the work the engine is doing.					
Weather	Load	Position of Radiator Curtain	Position of Air Valve	Position of By Pass Levers	Water to Carburetor
	Light	Adjust curtain just below boiling point 180-190° F.	Backward	Straight up	Off
Hot	Moderate	" " "	Forward	Set at about 45 degrees	Use water to prevent pinging if needed
	Heavy	" " "	Forward	Horizontal	Use sufficient water to prevent pinging
	Light	" " "	Backward	Straight up	Off
Moderate	Moderate	" " "	Forward	At about 45 degrees	Off
	Heavy	" " "	Forward	Horizontal	Off unless working very heavy
	Light	" " "	Backward	Straight up	Off
Cold	Moderate	" " "	Backward	At about 45 degrees	Off
	Heavy	" " "	Backward	Horizontal	Use water if needed to prevent pinging

Regulation of the water valve, exhaust manifold heat control, radiator curtain, and air damper on the Case Air Washer were subjective judgments left to the farmer. The operator manual provided general guidelines.

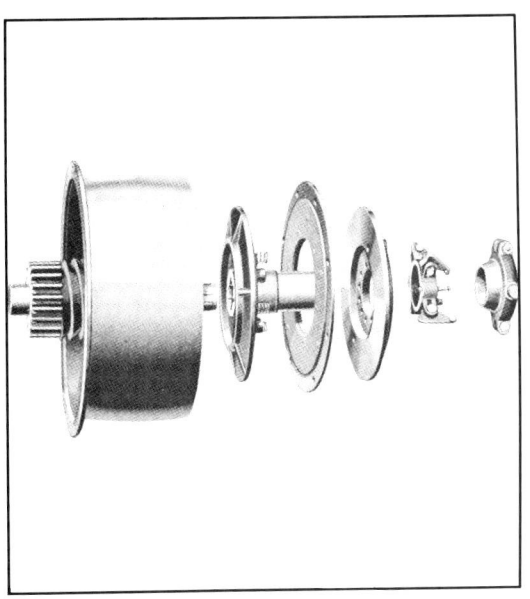

Illustration of clutch from *Better Farming with Better Tractors*. The Model 12–20 introduced the Twin Disk clutch system to the Case line.

The crossmotor tractors featured a fully enclosed differential and final drive.

suction was sufficient to draw water into the carburetor and made starting difficult.

These various features required careful attention on the part of the operator. Detailed instructions in the operator manual included directions explaining how air temperature and work load influenced various settings. Yet, as stated in the manual, the system relied entirely on the farmer's judgment.

The two-speed transmission and final drive characteristics of the 12–20 did not change from those of the 10–18. Case did, however, replace the expanding-shoe friction clutch with an improved Twin Disk clutch system. Forward speeds were 2¹/₅ and 3mph.

The 12–20 was also available in an industrial version. It differed slightly from the agricultural model. The Case Air Wash filter was replaced by a screen-type air cleaner, and the 12–20 industrial burned gasoline only. A seat with backrest was available.

Rear wheels were hollow cast-iron disks, onto which solid rubber tires were pressed. Each wheel could be filled with up to 300lb of sand for added weight. Front wheels were solid cast iron and were also fitted with rubber tires. The industrial tractor weighed

5,200lb without sand. Optional gearing raised the forward speeds to 3 and 4¹/₄mph.

The Model 12–20 was priced at $985 in 1925. The industrial version was priced at $1,385.

Model 15–27; Model 18–32 and Model K

The Model 15–27, Model 18–32, and Model K (the 1928 version of the 18–32)

Model 15–27 Specifications

Recommended for four 14in plows under ordinary conditions, a 10ft double disc harrow, and a 28x46in thresher with attachments

Rated Belt Horsepower: 27
Rated Drawbar Horsepower: 15
Rated Drawbar Pull: 2,500lb
Maximum Belt Horsepower: 33
Maximum Drawbar Horsepower: 21–24
Cylinders: Four: bore, 4¹/₂in; stroke, 6in
Normal Engine Speed: 900rpm
Ignition: Berling magneto with impulse coupling
Carburetor: Kingston vertical, with single nozzle; size, 1³/₈in
Fuel Capacity: Gasoline, 2³/₄gal; kerosene, 20gal
Cooling System: Capacity, 11gal
Road Speed: 2¹/₄mph and 3mph
Overall Length: 127in
Overall Width: 72in
Overall Height (without exhaust pipe): 68in
Wheelbase: 76¹/₂in
Shipping Weight: 6,350lb
Drawbar Height: 14in
Turning Circle: 27¹/₄ft
Front Wheels: Diameter, 32in; width, 6in
Rear Wheels: Diameter, 52in; width, 14in

This Model 12–20 industrial tractor (serial number 67288) featured optional seat with backrest, and hollow cast-iron rear wheels. Each wheel held up to 300lb of sand. *Case Company*

Illustration of a Model 15–27. Case rated the 15–27 as a three- to four-plow tractor. *Shields Library, Special Collections, University of California, Davis*

comprised the best-selling series of Case crossmotors. Essentially the same tractor, the 15–27 and 18–32 differed in horsepower output due to engine modifications.

Engine, frame, transmission, and final drive designs were similar to those of the smaller Case crossmotors: a crossmounted, vertical four-cylinder engine with 4$\frac{1}{2}$in bore and 6in stroke; a three-bearing crankshaft;

Model 15–27 Production Totals by Model Year
Source: J. I. Case Company

1919	1920	1921	1922	1923	1924	Total
5600	7500	17	1500	1311	1700	17628

Rear view of a Model 15–27.

Case claimed that the rigid frame of the Model 15–27 "assured permanent alignment of all bearings and perfect mesh of gears."

Case ad for the Model 15–27.

Model 15–27 with hard-rubber Kelly Springfield tires. Case promoted all sizes of the crossmotor tractors for use in industrial plants and road maintenance. *Case Company*

renewable cylinder sleeves; exhaust heated manifold; water air cleaner (some later units of the 18–32 were fitted with a Donaldson-Simplex air cleaner); carburetor with water-feed valve; twin disk clutch fitted inside the belt pulley; cast-iron frame; and fully enclosed, all spur gear, two-speed transmission, and final drives.

The 15–27, built from 1919–1924, was fitted with a Kingston carburetor and Berling magneto. Its centrifugal ball-type governor maintained a rated 900rpm engine speed. In its Nebraska test of April 1920 (Official Tractor Test No. 04), the 15–27 developed 31.23 maximum belt hp. In low gear, it developed 18.80 maximum drawbar hp and maximum drawbar pull of 3,440lb.

The 18–32, built from 1925–1927, and the K, built in 1928, were fitted with a Kingston carburetor and Bosch magneto. Engine speed was 1000rpm, the critical difference between the 15–27 and 18–32.

The 18–32 was tested at Nebraska in October 1924 (Official Tractor Test No. 109). Its engine developed 36.73 maximum belt hp, 24.01 maximum drawbar hp, and maximum drawbar pull of 3,882lb.

The Model 15–27 was priced at $1,680 in 1921. The industrial version was priced at $2,150.

The production total of seventeen tractors in 1921 reflected the depth of the post-World War I recession, which hit the agricultural sector of the economy in 1920 and 1921. The war years had seen total cash farm income soar from $6 billion in 1914 to nearly $14.5 billion in 1919. With this growth, annual demand for tractors had increased to a level of 136,162 units in 1919. In 1920 sales grew by more than 19 percent to a level of 162,988 tractors.

The economic growth spurred by World War I was suddenly and drastically reversed in 1920. Wholesale farm prices fell sharply.

Model 18–32 photographed at Rollag.

Model 15–27 photographed at Rollag.

Model 18–32 Specifications

Rated Belt Horsepower: 32
Rated Drawbar Horsepower: 18
Maximum Belt Horsepower: 36
Maximum Drawbar Horsepower: 24
Cylinders: Four: bore, 4½in; stroke, 6in
Normal Engine Speed: 1000rpm
Ignition: Bosch magneto with impulse coupling; Spark plugs: 7/8in US standard thread
Carburetor: Kingston vertical, with single nozzle; size, 1 3/8in
Fuel Capacity: Gasoline, 2 3/4gal; kerosene, 20gal
Cooling System: Capacity, 11gal
Road Speed: 2½mph and 3⅓mph
Overall Length: 127in
Overall Width: 72in
Overall Height (without exhaust pipe): 68in
Wheelbase: 76½in
Shipping Weight: 6,225lb
Drawbar Height: 14in
Turning Radius: 13½ft
Front Wheels: Diameter, 32in; width, 6in
Rear Wheels: Diameter, 52in; width, 14in

Cash farm income slipped to $12.5 billion in 1920 and plunged to $8.1 billion in 1921—a 44 percent decline from 1919.

The collapse of the farm economy had dire consequences for the farm implement industry. In 1921, tractor production fell by 64 percent. Losses were heavy, and many man-

Model 18–32 and Model K Production Totals by Model Year

Source: J. I. Case Company

1925	1926	1927	1928	Total
2300	3550	4000	5000*	14850

*Model K Production, 1928 only. Note that later units of the Model K featured wheels with rolled edges (earlier wheel rims were flat), operator platform, and fenders that closely resembled those of the Model L

The Model 22–40 weighed over 9,200lb, with maximum drawbar pull of 4,965lb. Case rated it as a five-plow tractor. *Shields Library, Special Collections, University of California, Davis*

ufacturers were forced into mergers or bankruptcy. Case, of course, survived this tough period. However, total Case tractor production in 1921 was just over 200 units.

By 1924, the farm economy had improved. However, as can be seen in the production figures below, annual demand for the 18–32 never equalled that of the early Model 15–27.

The Model 18–32 was priced at $1,350 in 1927. The industrial version was priced at $1,520.

Model 22–40; Model 25–45 and Model T

According to Case sales literature, the Model 22–40 was "especially suited to the requirements of the farmer with large acreage, or one who [had] a great deal of belt work to do." Case rated the tractor as capable of handling five 14in plow bottoms, two 8ft tandem disk harrows, or a 32x54in thresher. Introduced in 1919, the 22–40 was built through 1924. In 1925, following a second Nebraska Test that demonstrated its increased performance, the tractor was redesignated the Model 25–45. In 1928, the 25–45 became the Model T.

Model 22–40 Specifications
Recommended for five 14in plows under ordinary conditions, two 8ft tandem disc harrows, and a 32x54 thresher with attachments

Rated Belt Horsepower: 40
Rated Drawbar Horsepower: 22
Rated Drawbar Pull: 3,760lb
Maximum Belt Horsepower: 45–48
Maximum Drawbar Pull: 4250-4750lb
Cylinders: Four; bore, 5½in; stroke, 6¾in
Normal Engine Speed: 850rpm
Ignition: Bosch magneto with impulse coupling
Carburetor: Kingston vertical, with single nozzle; size, 2in
Fuel Capacity: Gasoline, 3¾gal; kerosene, 26½gal
Cooling System: Capacity, 15½gal
Road Speed: 2⅕mph and 3⅕mph
Overall Length: 153in
Overall Width: 82½in
Overall Height: 90in
Wheelbase: 96in
Shipping Weight: 9,200lb
Drawbar Height: 17in
Turning Radius: 20¼ft
Front Wheels: Diameter, 40in; width, 8in
Rear Wheels: Diameter, 56in; width, 16in; optional 10in extension rims

Make MORE Money with a CASE Tractor

THE Case 22-40 pulls four or five 14-inch plows; one 10-foot tandem disk; 9-foot tandem disk with 3-section spike tooth harrow; 10-foot tandem disk and 20-shoe grain drill; 8-foot binder and 8-foot tandem disk; two 8-foot binders, 10-foot road grader. In the belt it drives a Case 32x54 thresher with all attachments; a Case 20-inch silo filler, etc.

Any farmer whose work is heavy enough to require this much power, cannot make a mistake in owning a Case 22-40. No more dependable tractor was ever offered for heavy field and belt work.

Let us show you what this tractor can do for you in the way of making more money.

Case ad 4-2407 promising owners they would "Make More Money" with a Model 22–40.

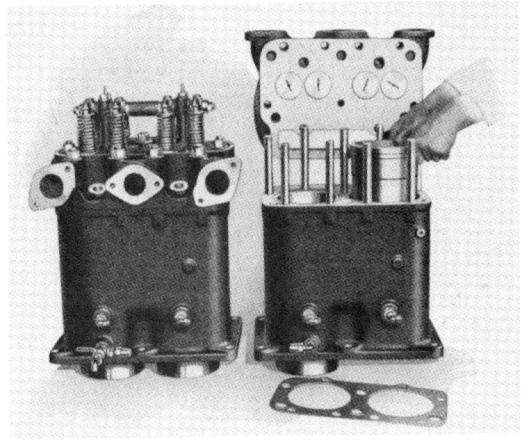

The Model 22–40 engine featured four-cylinders cast in pairs with individual heads.

Rear view of the Model 22–40. Standard rear wheels were 16in wide.

The 22–30 weighed approximately 9,220lb and required a substantial frame to support its weight. Rather than cast as one piece, the frame was built up from 7 and 8in channel steel and boiler plate.

The crossmounted vertical four-cylinder engine featured cylinders cast in pairs, with removable or renewable cylinder sleeves. The cylinder heads were also cast in pairs. Cylinder bore measured 5½in; stroke measured 6¾in. The crankshaft was carried on three main bearings. The Case-built centrifugal ball-type governor controlled engine speed at 850rpm.

The engine was equipped with a Kingston Type E, single-nozzle, 2in carburetor; a more simplified manifold with cold air damper for controlling heat; and Bosch ZR4 magneto.

The 22–40 expanding-shoe-type clutch was fitted inside the belt pulley. Activated by a hand lever, it served both the transmission and belt pulley.

The transmission gears were carried in a cast-iron case, bolted to the frame. It featured two forward speeds, with selection through sliding gears. As with the smaller crossmotor

Model 22–40 Production Totals by Model Year
Source: J. I. Case Company

1919	1920	1921	1922	1923	1924	Total
2	1020	181	100	297	250	1850

Model 25–45 and Model T Production Totals by Model Year
Source: J. I. Case Company

1925	1926	1927	1928	Total
115	340	375	150*	980

*Model T production

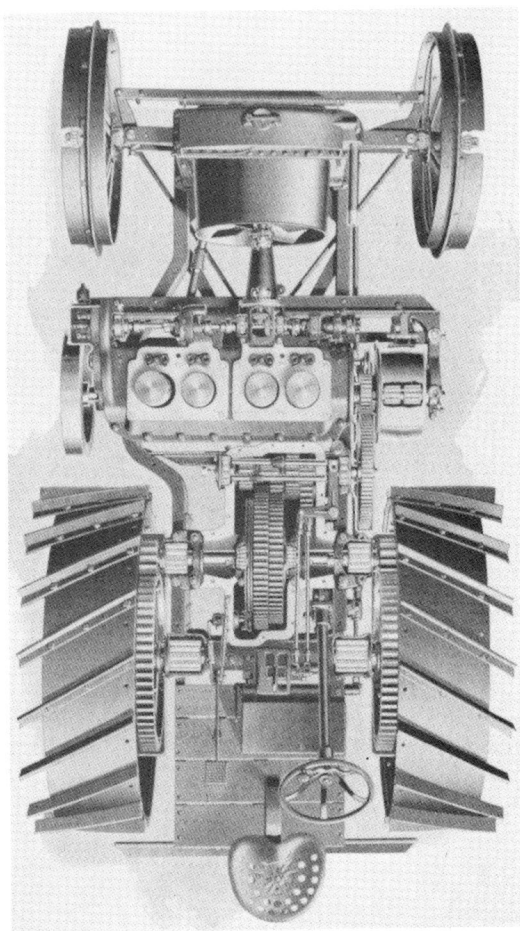

The Models 22–40 and 25–45 featured the same simple spur gear transmission and final drive designs as did smaller crossmotor models.

models, power was transmitted exclusively through spur gears. The final drive system was fully enclosed.

In its Nebraska Test of April 1920 (Official Tractor Test No. 04), the 22–40 developed 40.16 belt hp under rated load, a maximum 29.04 drawbar hp, and maximum drawbar pull of 4,965lb. The Model 22–40 was priced at $3,100 in 1921.

The Model 25–45 and Model T were nearly identical to the 22–40. A careful comparison of published specifications reveals no differences other than a change in the model of Kingston carburetor fitted to the engine.

Yet, the results of the November 1924 Nebraska test of the 25–45 (Official Tractor Test No. 109) indicate a significant improvement in performance over that of the 22–40: 45.18 belt hp under rated load, a maximum 32.96 drawbar hp, and maximum drawbar pull of 5,750lb.

Total production of the 22–40, 25–45, and T equalled only 2,468 units, considerably beneath that of the smaller crossmotor models. This again reflected the more limited demand that existed for tractors of this size. Priced at $3,100 to $3,500, these were expensive machines that only the wealthiest farmers could justify purchasing and operating.

Model 40–72

The largest of Case crossmotors, the Model 40–72 was an impractical tractor perhaps better suited to road building and

Model 40–72 Production Totals by Model Year
Source: J. I. Case Company

1920	1921	1922	1923	Total
1	—	20	20	41

The Model 25–45 weighed in at 10,035lb.

Model 22–40 at the 1992 WMSTR event.

construction work than to farming. Impractical or not, the 40–72 was an imposing machine of gigantic proportions. Its overall height exceeded 9ft, and it weighed nearly 10 tons. With only forty-one units built between 1920 and 1922, and as few as five units extant, it is the rarest of Case crossmotors.

The design and construction of the 40–72 closely paralleled that of the 22–40: the tractors weight was carried on a frame built up of 8in and 10in channel steel and boiler plate; its crossmounted four-cylinder engine featured cylinders cast in pairs, with removable cylinder sleeves; separate cylinder heads; and three-bearing crankshaft. Cylinder bore measured 7in; stroke measured 8in.

The engine featured a Kingston Model L, single-nozzle, 2½in carburetor; water valve; manifold with cold air damper and exhaust bypass; Bosch ZR4 magneto and Case-built centrifugal ball-type governor. Engine speed was 750rpm.

The 40–72 employed a twin disk-type clutch, fitted inside the belt pulley. Transmission gears were carried in a cast-iron case. The transmission featured two forward speeds, with selection through sliding gears.

Model 40–72 Specifications

Recommended for eight to twelve 14in plows under ordinary conditions, three 10ft tandem disc harrows, and a 40x62 thresher with attachments

Rated Belt Horsepower: 72
Rated Drawbar Horsepower: 40
Rated Drawbar Pull: 7,250lb
Maximum Belt Horsepower: 90–95
Maximum Drawbar Horsepower: 66
Cylinders: Four; bore, 7in; stroke, 8in
Normal Engine Speed: 750rpm
Ignition: Bosch magneto with impulse coupling
Carburetor: Kingston vertical, with single nozzle; size, 2½in
Fuel Capacity: Gasoline, 9gal; kerosene, 52gal
Road Speed: 2.07mph and 2.97mph
Overall Length: 200in
Overall Width: 105in
Overall Height: 110in
Wheelbase: 124in
Shipping Weight: 21,200lb
Drawbar Height: 20in
Rear Wheels: Diameter, 72in; width, 20in; optional 12in extension rims

Power was transmitted through spur gears, in the same manner as the smaller crossmotor tractors. Unlike the smaller units, the 40–72 offered a means to lock the differential. As a

Illustration of the Model 40–72 from *Road and Farm Work.*

Weighing nearly 10 tons and with overall height exceeding 9ft, the Model 40–72 is an impressive sight. Two 40–72s appeared at the 1992 WMSTR event.

result, power was transmitted to only one wheel.

In its Nebraska test of April 1923 (Official Tractor Test No. 090), the 40–72 developed a phenomenal 91.42 belt hp under maximum load, 49.87 maximum drawbar hp, and an earth-moving 10,868lb maximum drawbar pull.

Case rated the 40–72 as capable of handling eight to twelve 14in plow bottoms or, in road work, a 12ft grader. The Model 40–72 was priced at $4,000 in 1923.

Industrial, Road, Orchard, and Rice Tractors

In addition to standard farm models, the crossmotor tractors were offered for a variety of applications that included industrial and road work, orchards, and rice fields.

The industrial version of the Model 12–20 was reviewed above. The Model 10–18 and Model 15–27 were also offered in industrial versions. It would appear that these tractors varied little from the farm versions. They were marketed primarily to municipalities, lumber yards, manufacturing plants, foundries, and steel plants for use in hauling. Equipped with rubber tires, vibration was reduced from that of the steel-wheeled farm tractor and damage to paved surfaces was eliminated.

Case marketed the 12–20, 15–27/18–32, 22–40/25–45, and 40–72 units for road work, with few apparent modifications. Optional cast-iron road wheels equipped with heavy road cleats were available for the Models 15–27/18–32 and 22–40/25–45. The wheels added 1,065lb to the weight of the 15–27/18–32 and approximately 3,200lb to the weight of the 22–40/25–45.

Case rated the road tractors based on their standard performance features, but translated into terms relevant to the work carried out by the contractor or highway department official:

The 12–20 was rated to pull a 6 or 7ft grader or handle two 1 yard wheel scrapers; its belt would operate a 9 ton/hour rock crusher. Case literature stated, "the tractor is especially well adapted to road maintenance and patrol. In those sections where systematic patrol work is carried on, one man with a

The flywheel and clutch mechanism of the Model
40–72.

A May 1920 photograph of the first Model 40–72
built. It handled an eight-bottom plow with ease.
Case Company

Model 22–40 equipped with steel wheels with road cleats.

12–20 tractor can take care of approximately 13 miles."

The 15–27/18–32 was rated to pull a 7 or 8ft grader; three 1 yard wheel scrapers; its belt would operate a 12–16 ton/hour rock crusher; and it would permit one worker to patrol 20 to 30 miles per day.

The larger 22–40/25–45 tractors were rated to pull a 10 to 12ft grader, four to five 1 yard scrapers, a 20 to 30ft spread maintainer, or a 25ft road planer. With drawbar pull of 3,760lb at 2.2mph, the 22–40 was capable of pulling 7 to 10 ton trailer loads.

The 40–72 was simply promoted as having "as great a capacity for work as we believe practical to build in one unit." What more could be said about a tractor with 12,000lb drawbar pull and a maximum 90hp at the belt pulley?

Undoubtedly, Case sold a number of tractors for applications other than farming. The exact number is difficult to determine, as the tractors were pulled from the standard production line.

Orchard versions of the 12–20, 15–27, and 18–32 models were also available. Case promoted them as ideal for work in orchards, as their high clearance but low centers of gravity allowed stability on sloping or rough ground.

Special orchard equipment available from the factory or for field installation included the following: orchard exhaust, a low-profile system with exhaust pipe horizontal to the manifold; citrus fenders; high, chilled-face cast-iron spade lugs; 3in skid rings for front wheels, which could be bolted on the regular skid rings; radiator screen; and optional oil-type air cleaner, in place of the Case Air Wash filter.

"Rice Field Specials" were specially equipped versions of the 12–20, 15–27/18–32, and 22–40/25–45 tractors suited to work in rice fields. These were standard tractors equipped as follows: oil-seal differential; water air washer with periscope pipe, an upright intake pipe; special angle-iron lugs, 24in or 30in long for the 12–20, 24in or 33in for

The Models 12–20, 15–27, 18–3, 22–40, and 25–45 Rice Field Specials were equipped with extra-long angle-iron lugs, periscope intake pipe, and oil seal differentials. All but the 22–40 and 25–45 were fitted with a power takeoff.

the 15–27/18–32, and 36in for the 22–40/25–45 tractors; and power takeoff for the 12–20 and 18–32.

Options, Attachments, and Accessories

The Case optional power takeoff was available either from the factory or for field

The Axle Extension Furrow Guide was designed to aid steering and keep the tractor level.

installation to fit any model 12–20 or 18–32. The PTO unit was powered by a spur gear driven off the transmission and was designed to be removed from the tractor when not in use. The splined end measured $1^{5}/_{16}$in in diameter and rotated at 516rpm on the 12–20, and at 538rpm on the 18–32.

Other options included a radiator screen, belt (pulley) guide, adjustable safety spring hitch, acetylene headlight, three-tone chime exhaust whistle, canvas cover, canopy (15–27/18–32 only), and 250gal to 380gal fuel tender wagons.

Of particular interest were the Case Axle Extension Furrow Guide, Case Tractor

Model 10–18 with optional Extension Control. *Case Company*

Guide, and Case Extension Control. The Case Axle Extension Furrow Guide (for 12–20 and 15–27/18–32 tractors) consisted of a front axle extension and wheel that were attached to the yoke of the front axle in place of the standard front wheel. This allowed the front wheel to drop into the furrow at the same time the tractor maintained its horizontal position.

The Case Tractor Guide consisted of a bent pipe, one end of which pivoted on an extension shaft bolted to the hub of the front wheel, while the other end, which was curved, ran against the bottom and side of the furrow. Case claimed the device eliminated "the continuous attention necessary on the part of the operator when plowing." The Case Tractor guide was available for all but the 40–72 tractor.

The Case Extension Control permitted the operator to steer and control a tractor from the seat of a drawn implement, such as a mower, binder, or wagon. The beveled gears of this telescoping steering shaft meshed to

Case Factory Crossmotor Standard and Optional Wheel Equipment

Model	Std. Rear Wheel Equipment	Optional Rear Wheel Equipment	Optional Front Wheel Equipment
10–18	Angle-iron grouters	Spade lugs, 8 and 12in extension rims	
12–20	Angle-iron grouters (before 1925)	Spade lugs, inverted angle-iron and tie ring, road cleats, 6in extension rims, rubber tires, 24 and 30in long, deep angle-iron grouter for rice tractors	4in extension rims; extension front axle; rubber tires
	Spade lugs (from 1925)	Angle-iron grouters, sand lugs, road cleats, 6in extension rims, rubber tires, 24 and 30in long, deep angle-iron grouter for rice tractors	4in extension rims; extension front axle; rubber tires; high skid rings
15–27	Angle-iron grouters	Spade lugs, inverted angle-iron and tie ring, road cleats, 4, 6, and 8in extension rims, rubber tires, cast-iron road wheels, 24 and 33in long, deep angle-iron grouter for rice tractor	4in extension rims; extension front axle; rubber tires
18–32	Spade Lugs	Angle-iron grouters, cast-iron road wheels, inverted angle-iron and tie ring, road cleats, sand lugs, 4, 6, and 8in extension rims, rubber tires, cast-iron road wheels, 24 and 33in long, deep angle-iron grouter for rice tractor	4in extension rims; extension front axle; rubber tires; high skid rings
22–40	Angle-iron grouters	Spade lugs, road cleats, cast-iron road wheels, 10in extension rim, 36in long, deep angle-iron grouter for rice tractor	
25–45	Angle-iron grouters	Spade lugs, road cleats, cast-iron road wheels, 10in extension rim, 36in long, deep angle-iron grouter for rice tractor	
40–72	Angle-iron grouters	12in extension rims	

those on the steering column. Clutch and brake were handled by means of an endless rope, which Case claimed gave "perfect control at all times."

Wheel and Tire Equipment

Steel wheels were standard equipment on all crossmotor tractors. Options included the type of lug, cleat, or angle-iron grouter bolted to the rear wheel. Rubber tires, cast-iron road wheels, and front and rear wheel extension rims were also available.

A problem that faces many collectors is the question of wheel and tire equipment. Even for tractors built prior to when pneumatic tires were an option, it is sometimes difficult to know what was standard, optional, or cobbled up by a local blacksmith.

Collecting Comments

The Case crossmotors are among the more acclaimed and popular of Case tractors. Their distinctive, stocky design appeals to many collectors. The one-piece, cast-iron frame of the 9–18B, 10–18, 12–20, 15–27, and 18–32 was unique and adds to their present value. The larger 22–40 and 25–45 were low-production units, which makes them more attractive to collectors.

And the Model 40–72? Logic and all forms of common sense go out the window when confronted by such a tractor. There were only forty-one units built, and only five are known to exist today. On the basis of uniqueness and availability alone, it is the only true five-star crossmotor tractor.

While every antique tractor shows the effects of heavy usage and time—rust, wear, modified or missing parts—there are three particular points to consider regarding the crossmotor tractors. First, the Model 9–18 and earliest versions of the 10–18 suffered from a poor oiling system. Lubrication was by plunger pump to the main bearings, and by overflow and splash trays to the cylinders and other working parts, whereas later units of the 10–18 (and all later Case models) featured full pressure feed-type lubrication with a drilled crankshaft. The effect of poor lubrication is, of course, excessive wear.

Second, the crossmotor manifold was complex and fabricated of cast iron. It would be surprising to find an original manifold that had not cracked. When inspecting a crossmotor tractor, keep in mind that if water has seeped into the manifold it may also have seeped into the head.

Third, the Case Air Washer relied on water to filter out contaminants. Rust is their enemy; few such units remain in serviceable condition.

Industrial, rice, and orchard versions of the crossmotor tractors are more rare than the standard farm versions. Nevertheless, I would only place a premium on the value of an orchard tractor, and then only if equipped with full citrus fenders, which was an option.

Chapter 5

Standard and Row Crop Tractors 1929–1940

The L, C, and R Series

Rating	Model	Remarks
★★ ¹/₂	L Series	Recommended for three to four 14in plows

The Case crossmotor tractors had become rather dated by the mid–1920s. Their broad-framed stocky appearance, transverse-mounted engines, and two-speed transmissions had limited appeal to farmers who had been exposed to the unitized construction of the Fordson and the trim, sturdy design of the three-speed, general-purpose Farmall.

The Model L was the first Case tractor to employ a unitized design. *Shields Library, Special Collections, University of California, Davis*

These two tractors had greatly influenced the industry.

The Fordson dominated tractor sales between 1919 and 1926. By 1928, 750,000 units had been sold. While not without weaknesses, its compact, lightweight design revolutionized the industry.

The Fordson was built up from two stressed cast-iron components: a transmission and rear-axle housing cast in one piece and bolted to its four-cylinder cast engine block. A cast front axle and steering support mount was bolted to the engine block. The strength and rigidity of stressed cast iron permitted the tractor the first unitized—or made as one unit—structure without chassis.

Eventually, the Fordson unitized design became the industry norm. Not only did it eliminate the need for a separate frame, its three-piece, bolt-together construction facilitated assembly line manufacture—the keystone to mass-production and a lower cost of assembly.

In 1923, International Harvester introduced the Farmall, the first general-purpose tractor. Able to plow, plant, cultivate, and harvest any type of crop, the general purpose tractor significantly reduced the man-hours needed to produce row crops. While Farmall production was limited to fewer than 1,100

tractors through 1925, its revolutionary design had an immediate and significant impact on the market. All manufacturers reacted and initiated or accelerated development of their own all-purpose designs. By 1931, at least a dozen general-purpose tractors were on the market.

Sales of general-purpose tractors proved strong, as farmers welcomed the economy and versatility these tractors offered. By 1930, IH had built the 100,000th Farmall, and by 1940, general-purpose tractors comprised 90 percent of the 249,397 wheel tractors manufactured in the United States.

Case began experimenting with new tractor designs to replace the crossmotor series as early as 1925. Two new tractors emerged from this work: the Model L, introduced in February 1929; followed later in the year by the Model C (and its row-crop derivative Model CC). These tractors featured a unitized design, four-cylinder Case-built engine, three-speed transmission, and roller-chain final drive.

L Series

The Model L was a well-proportioned and well-conceived standard tractor. Its fun-

The Model L featured full fenders and protective sheet metal, which kept dust down and away from the operator.

The Model L offered a trim appearance and high clearance under the front axle, a contrast from the stocky crossmotor series.

damental engine design remained in production for twenty years; and the basics of its transmission were employed by Case into the 1960s.

A unitized design, its engine was bolted to the rear transmission/final drive casting. A separate casting carried the radiator and front axle and was bolted to the front of the engine.

The Model L's vertical four-cylinder engine was cast en bloc and featured removable cylinder sleeves. Bore measured $4^5/_8$in; stroke measured 6in. The three-bearing crankshaft measured $2^3/_4$in in diameter. Lubrication was force-fed through drilled holes in the crankshaft, with oil supplied under pressure by a geared-type pump.

The early Model L was fitted with a $1^1/_2$in, Kingston L–3 carburetor similar to that fitted to the crossmotor series. The manifold and exhaust were greatly simplified from that of the crossmotor series. Case touted its "special combination" manifold, with one simple heat adjustment, as effective at vaporization of either low- or high-grade fuels. A Case-built fly ball governor maintained rated en-

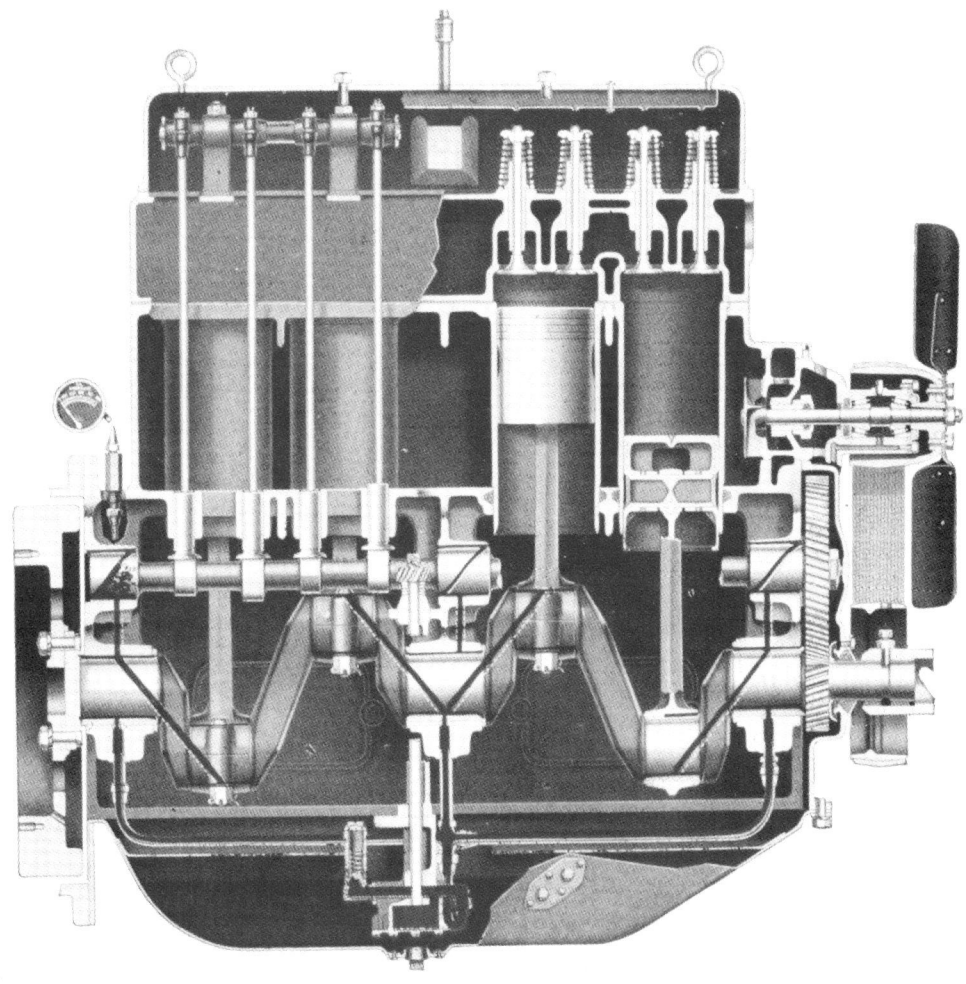

The Model L four-cylinder engine (and the smaller version fitted to the C Series) remained virtually unchanged for two decades.

The Case-built fly ball governor, as fitted to the Model L.

gine speed of 1100rpm, and a Bosch FU–4-BR magneto with impulse coupling provided spark.

At serial number 301843, the carburetor was changed to a Kingston model L–3L. The governor was also replaced. Eventually, Case went to a Zenith model K6A carburetor.

The L featured a Case-designed oil-bath air cleaner. Incoming air was pulled through a baffled screen and oil pool, through a mist of oil, and finally through oil-coated screens before it was drawn into the carburetor. The lower half of the air cleaner was removed for servicing—old oil was discarded and replaced by clean engine oil.

The Model L cooling system employed an 18in fan, driven by a 2½in flat belt. Water was circulated by a centrifugal pump. In 1934, at serial number 300723, a V-belt replaced the flat belt, with corresponding changes in pulleys on the water pump and crankshaft. The radiator was protected by a removable radiator screen that incorporated an adjustable canvas curtain. The curtain could be raised to cover all or any portion of the

The Case oil bath air cleaner. Case recommended that the unit be serviced daily.

The Model L water pump and fan. The 18in fan was driven by a 2½in flat belt on early units. In 1934, a V-belt replaced the flat belt.

radiator, as a means of maintaining proper engine temperature during cold weather.

The twin disk clutch assembly, steering gear, transmission, and final drive were enclosed in a one-piece casting. The transmission offered three forward speeds: 2½, 3¼, and 4mph. The clutch was hand operated.

Case moved from an all spur gear transmission and final drive on the crossmotors to a chain and sprocket final drive on the Model L. Case employed similar chain and sprocket final drives on most of its tractors into the 1950s.

Case cited five advantages to its hardened-steel, two-roller-chain final drive: less wear on sprockets than on gears, as the wrapping of chain around a sprocket brought more teeth into contact to carry the load; the pull on the chains was in the same direction as the push of the drive wheels, which tended to keep the axles in alignment, while a spur gear transmission tended to angle the axles; chains, being more flexible, absorbed shocks of starting and sudden

One-piece transmission and final drive housing of the Model L.

loads; chains, when lubricated and operated on the correct form of sprockets, proved highly efficient in transmitting power—less power was lost through friction; and chain

The Model L three-speed transmission and roller-chain final drive.

Model L and four-bottom Model B plow. *Case Company*

Model L and Model P combine. *Case Company*

drives gave superior service even if exposed to dirt.

A belt pulley was standard, as were fenders, a swinging seat, and set of tools. An optional power takeoff with 1³⁄₈in splined shaft was offered. It operated at 550rpm.

Other optional equipment included the following: a motometer, a water temperature gauge that could be mounted to the radiator cap; chime whistle; steel and wood-constructed winter cab; khaki canvas summer canopy; electric lights and 50 watt 6 volt generator; extension fenders used with rear extension rims; full-skirted "orchard" fenders; high air intake pipe; hood sides; and a spark-arresting muffler. Beginning in 1938 Case offered optional electric starter and lights for the Model L (serial number 421024 was the first unit to feature a flywheel ring gear). While this was a common option for industrial versions of the Model L (see below), fewer ag versions were fitted with electric starter.

Steel wheels were standard on the early Model L. Front wheels measured 30x6in and

The 30inx6in front wheel fitted to the Model L came standard with a 1³⁄₄in skid ring.

were fitted with 1³⁄₄in skid rings; rear wheels measured 48x12in and were fitted with 5in spade lugs. Optional wheel equipment included 6in cast-iron spade lugs; chilled sand

Model L with optional factory cab.

L Series Production Totals by Model Year*
Source: J. I. Case Company

1929	1930	1931	1932	1933	1934
6800	6361	3354	300	204	589

1935	1936	1937	1938	1939	1940	Total
2099	3266	4347	3630	28	700	31678

*Total of ag tractors only. An additional 2,136 industrial and other limited versions of the L were built

lugs; 2½in high, 21½in long angle-iron grouters; 2x1x13in road grouters; road bands; wheel scrapers, 6 and 8in rear extension rims; 3in high (front wheel) skid rings; and 2in wide (front wheel) guide rings.

In 1934, Case offered rubber tires with Case-made wheels and optional wheel weights. Front tire options included 7.50x16 and 7.50x18in tires. Rear options included 12.75x28, 13.50x28, and 9.00x36in tires.

Because rubber tires made it possible to drive on paved roads, farmers were now able to haul wagon loads or move from field to farmhouse more easily and at higher speeds. Case offered an optional high-speed field attachment that included axle sprockets, high-speed gear and pinion, and longer roller chains. This boosted maximum road speed, in third gear, to almost 12mph.

The Model L was not tested at Nebraska before October 1938 (Official Tractor Test No. 309). Its engine was rated at a maximum 47.04 brake hp. Tested on rubber tires, it developed 40.80 drawbar hp. Tested on steel wheels, it developed 31.94 drawbar hp. Maximum drawbar pull was 4,474lb on rubber tires; 4,472lb on steel wheels.

Case also built the Model L in industrial and military versions. Essentially identical to the farm tractor, the LI offered the following:

A nicely restored Model L at the 1992 WMSTR event.

a foot-operated clutch; hand and foot throttles; leaf spring-mounted, heavy-duty front axle; padded seat; and either solid or pneumatic tires. Initially offered with a three-speed transmission, an optional four-speed transmission was added.

Additional LI options included: belt pulley; PTO; automatic couplers; front and rear bumpers; no generator-type electric lights; generator, starter and electric lights; canopy; cab, hood sides; muffler; radiator curtain and screen; radiator guard; wheel weights; odometer, and more.

The Model L was priced at $1,295 in 1930; $1,175 on steel and $1,405 on rubber in 1934. The Model LI was priced at $1,695 in 1930.

Collecting Comments

The Model L was the first Case tractor built with a unitized design. In my opinion, it was one of the best-looking standard tractors of its day.

However, the L does not seem to command as high a price as we Case lovers might like or think it should. With little to distinguish early units from later units, the more desirable Ls would be those on steel wheels or with more rare factory options, such as a cab, hood sides, or full orchard fenders.

Take note of the following two things when inspecting the Model L: First, units prior to serial number 300723 featured a flat fan belt; units after that serial number fea-

Case introduced the industrial Model LI in 1930. Offered with three-speed transmission, a four-speed unit was optional. *Case Company*

tured a V-belt. Pulleys on the water pump and crankshaft corresponded accordingly. At some point, replacement water pumps were sold with V-pulleys only, which would have forced conversion. Consequently, few of the flat belt units have survived.

Second, check the manifolds and heads carefully, particularly on a tractor not running; both are prone to cracking. Many Model Ls will have had a replacement Model LA manifold or head fitted.

C Series

From its four-cylinder engine to its three-speed transmission and roller chain final drives, the Model C was virtually identical to the Model L but built on a smaller scale. While the Model L was rated as a three- to four-plow tractor, the Model C fell in the two- to three-plow range.

The C Series four-cylinder engine, with bore of 3⅞in and stroke of 5½in, developed 27 brake hp and 17 drawbar hp at 1100rpm under rated load.

The C Series was offered in a wide range of models: a standard tractor, Model C; an industrial version, Model CI; a row-crop tractor, Model CC; orchard and vineyard tractors, Model CO and Model CO Vineyard Special; a high-clearance tractor, Model CH; and a sugar cane special, Model CCS.

The general-purpose Model CC proved the most popular. Case offered the Model CC–3, in both a one- or two-wheel tricycle front version (referred to as a three-wheel model), and a standard front axle Model CC–4.

The Model CC featured the type of options that made the row-crop tractor more popular than the standard tractor: greater

A 1938 Model C. Design of this two- to three-plow tractor was virtually identical to that of the larger Model L. *Case Company*

Rating	Model	Remarks
★★	C	Standard tractor, approximately 40% of C Series total production
★★	CC–3	Better than 50% of C Series total production
★★	CC–4	Built in limited numbers
★★	CI	The industrial version of the standard Model C
★★★	CO	Equipped with cast front wheel, steel rears with extension rims, and extension fenders ideal. Hood sides are rare
★★★	CO-VS	Narrow rear vineyard special. Only 190 built
★★★	CCS	Just over 100 units built in 1937 and 1938
★★★★	CH	Seven built in 1938

ground clearance, which permitted travel over young crops; adjustable rear tread width to accommodate operation in any type of row crop; and the addition of independent differential foot brakes, which shortened the turning radius and made faster turns possible at the headlands.

Among the most distinctive features of the CC were its "goose neck" front wheel support; its long steering arm mechanism, often referred to as a "chicken roost"; and the method by which its rear tread width was varied. Rather than employ long rear axles on which wheels were moved in or out to adjust tread width, Case employed extension spools of varying size. Spools were bolted to

either side of the axle on the same flange that supported the wheel in its narrowest tread width (48in). By using 2, 10, or 12in spools and by reversing the wheels, the CC rear tread width was adjustable in 4in intervals from 48in (no spools, wheels in) to 84in (12in spools, wheels out).

In 1935, Case introduced Motor Lift for the Model CC, a mechanical power lift feature that employed engine power to raise or lower implements. The lift was a simple worm-and-gear unit driven by the tractor power takeoff, and activated by a trip button mounted through the floor of the operator platform. Both the PTO and Motor Lift were optional features.

The Model C and a Case two-way plow. *Case Company*

A Model C at the Rollag show.

1931 Model C Specifications

Recommended for two to three 14in plows under ordinary conditions, or a 22x36 thresher with attachments

Rated Brake Horsepower: 27
Rated Drawbar Horsepower: 17
Maximum Brake Horsepower: 29.81
Maximum Drawbar Horsepower: 19.6
Cylinders: Four; bore, $3^7/_8$in; stroke, $5^1/_2$in
Normal Engine Speed: 1100rpm
Ignition: Bosch FU4ARS magneto with impulse coupling
Carburetor: Kingston L 3 vertical with single nozzle
Fuel Capacity: Gasoline, 2gal; kerosene, 18gal
Cooling System: Capacity, 5gal
Road Speed: $2^1/_3$mph, $3^1/_3$mph, and $4^1/_2$mph
Overall Length: 114in
Overall Width: 60in
Height to Top of Hood: 48in
Wheelbase: 66in
Total Weight: 4,155lb
Turning Radius: 10ft
Front Wheels: Diameter, 28in; width, 5in
Rear Wheels: Diameter, 42in; width, 12in

cluded the following: $5^3/_4$in or 6in spade lugs; 6x48in rear extension rim; steel cone lugs; narrow rear wheel with either spade or sand lugs; open or "southern" type rear wheel with spade lugs; and road bands.

Both Model C and Model CC were tested at Nebraska in 1929. In its test, the Model C (Official Tractor Test No. 167) developed 29.81 maximum brake hp, 19.60 maximum drawbar hp, and maximum drawbar pull of 3,289lb. The Model CC (Official Tractor Test No. 169) developed 28.97 maximum brake hp, 22.70 maximum drawbar hp, and maximum draw-bar pull of 2,950lb.

The Model CO orchard tractor was built especially low and compact to allow opera-

Standard wheels were 25x4in steel in the front, with $1^1/_2$in skid rings; 48x8in steel in the rear, with 4in spade lugs. Rubber tires were offered as optional equipment, beginning in 1934. Additional steel wheel options in-

C Series Production Totals By Model Year
Source: J. I. Case Company

Model	1929	1930	1931	1932	1933	1934	1935	1936	1937	1938	1939	1940	Total
C	1139	5420	1154	344	98	856	1925	3450	3900	2123	42	—	20451
CI	50	138	91	15	1	90	28	75	76	141	71	15	791
CC–3	48	4223	1590	430	151	856	3880	7049	6101	4290	38	—	28656
CC–4	—	—	453	—	—	—	5	101	175	434	—	—	1168
CO	148	250	168	—	—	181	54	200	161	113	—	—	1275
CO-VS	—	—	—	—	—	—	61	99	—	29	1	—	190
CCS	—	—	—	—	—	—	—	—	65	50	—	—	115
CH	—	—	—	—	—	—	—	—	7	—	—	—	7

tion among low-hanging limbs and fruit. Its full-skirted, downward-sloping fenders lifted low-hanging branches and protected both tree and fruit from damage. The solid front wheel disks and smooth sides of the tractor left nothing to rake limbs when operated close to trees.

The CO employed the same engine and transmission as were used on the CC. It also featured differential turning brakes, which were a benefit in the tight turns and loose soils of many orchards.

Rear tread width on the CO was fixed at 48in. For work in the more narrow rows of vineyards, Case offered the Model CO Vineyard Special. Its wheel tread was reduced to 38in. The Vineyard Special was built in limited numbers, beginning in 1935.

In 1937 and 1938, Case built a limited number of Model CCS (Case Cane Special) sugar cane tractors. With 30in ground clear-

Front view of the Model CC–3.

The 1930 Model CI "Golf Course Special." Case Company

Side view of the Model CC–3.

Side view of the Model CC–4.

1934 Model CC Specifications

Pulls a two- or four-row cultivator, or a two-bottom 14in plow. Operates a 22x36 thresher with all attachments

Rated Brake Horsepower: 27
Rated Drawbar Horsepower: 17
Maximum Brake Horsepower: 28.97
Maximum Drawbar Horsepower: 22.7
Cylinders: Four: bore, 3$\frac{7}{8}$in; stroke, 5$\frac{1}{2}$in
Normal Engine Speed: 1100rpm
Ignition: Bosch magneto with impulse coupling
Carburetor: Kingston vertical, with single nozzle
Fuel Capacity: Gasoline, 2gal; kerosene, 18gal
Cooling System: Capacity, 5gal
Road Speed: 2$\frac{2}{3}$mph, 3$\frac{3}{4}$mph, and 5mph
Height to Top of Hood: 53in
Wheelbase: CC 3, 89in; CC-4, 66in
Rear Tread: 48in, as shipped; widest spread 84in
Front Wheels: Diameter, 25in; width, 4in
Rear Wheels: Diameter, 48in; width, 8in

Front view of the Model CC–4.

ance and 70in tread, the CCS was built to clear cane ridges. The CCS featured standard PTO and Motor Lift. Its high-speed gearing permitted an 11mph road speed, in addition to operating speeds of 2$\frac{1}{4}$ and 3mph.

In 1938, Case built seven units of the high clearance Model CH. A modified standard Model C, the CH offered 14in ground clearance at the front axle, and 16$\frac{1}{2}$in ground clearance at the drawbar. Case advertised the tractor as "especially desirable for deep plowing, or working in soft, loose ground."

1936 Model CO Specifications

Recommended for two- or three-bottom plow, operates a 22x37 thresher with attachments

Rated Brake Horsepower: 27
Rated Drawbar Horsepower: 17
Cylinders: Four: bore, 3$\frac{7}{8}$in; stroke, 5$\frac{1}{2}$in
Normal Engine Speed: 1100rpm
Ignition: Bosch magneto with impulse coupling
Carburetor: Kingston vertical, with single nozzle
Fuel Capacity: Gasoline, 2gal; kerosene, 18gal
Cooling System: Capacity, 5gal
Road Speed: Equipped with standard steel wheels, 2$\frac{1}{3}$mph, 3$\frac{1}{3}$mph, 4$\frac{1}{2}$mph; equipped with rubber tires, 2$\frac{3}{4}$mph, 3$\frac{1}{3}$mph, 5$\frac{1}{2}$mph
Wheelbase: 66in
Height to Top of Hood: 48in
Outside Turning Radius: 10ft
Front Wheels: Diameter, 28in; width, 5in
Rear Wheels: Diameter, 42in; width, 12in

Case newspaper ad from 1931.

Model CC with optional single front wheel operating in a California lettuce field. *Shields Library, Special Collections, University of California, Davis*

The CH was available on a choice of steel wheels or rubber tires. Special road gears were offered that permitted top speeds of 12–12½mph. Other options included PTO, ex- tension rims, wheel weights, muffler, hood sides, canopy, cab, electric lights, and sponge rubber cushioned seat.

The Model C was priced at $985 in 1931; the Model CC–3 and Model CC–4 at $1,025. The Model C was priced at $1,158 on rubber in 1934; the Model CO on steel at $1,005. The

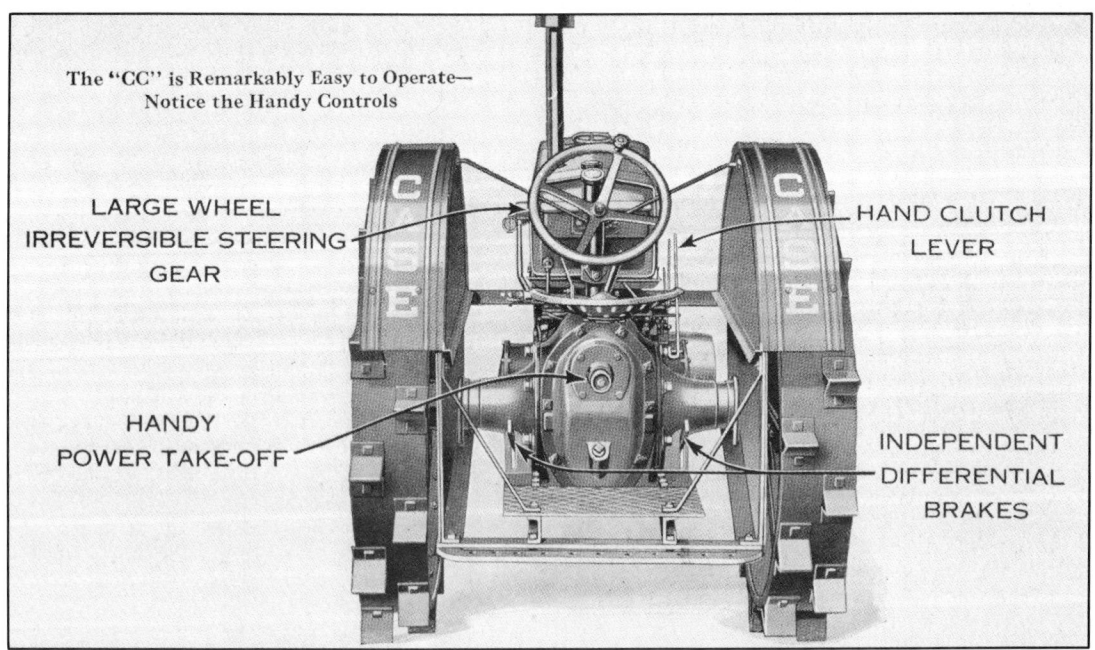

The "CC" is Remarkably Easy to Operate—
Notice the Handy Controls

LARGE WHEEL IRREVERSIBLE STEERING GEAR

HAND CLUTCH LEVER

HANDY POWER TAKE-OFF

INDEPENDENT DIFFERENTIAL BRAKES

The Model CC operator platform and controls. The CC featured hand-operated clutch and optional power takeoff. Independent differential brakes or "turning" brakes assisted in swinging the tractor around quickly at the end of a row.

Model CC and B−26 plow. *Case Company*

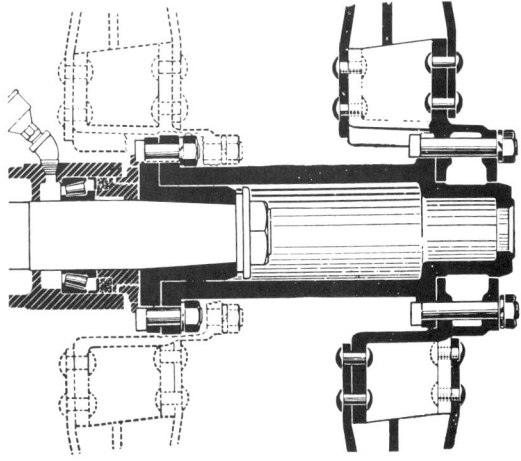

Model CC−3 with optional adjustable front axle. The long "chicken roost" steering arm and "goose neck" front wheel support became prominent features of the Case general-purpose tractor line.

The Model CC employed 2in, 10in, or 12in extension spools to adjust rear tread width. Rear tread width could be adjusted from 48 to 84in, in 4in increments.

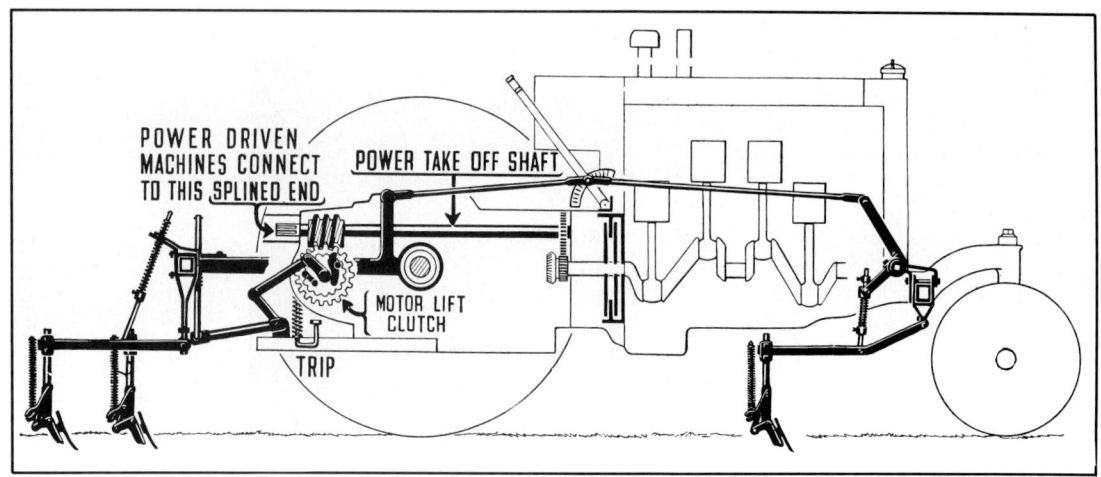

In 1935, Case introduced the Motor Lift, a mechanical power lift feature driven by the power takeoff.

Model CC–3 was priced at $1,095 on rubber in 1939. The Model C was priced at $900 on steel and $1,132 on rubber in 1940; the Model CH at $910 on steel and $1,100 on rubber.

Collecting Comments

The C Series, like the L Series, was a well-designed tractor. It has the advantage of being rarer than the later D Series; however,

Model CC and A–6 combine. The CC adapted easily to any type of drawbar work. *Case Company*

enough units were built so that spare parts are less of a problem than they are with the earlier crossmotor tractors.

When inspecting a tractor, buyers need to watch for the routine concerns: signs of head and manifold cracks or repaired cracks (many units will have been converted to D components); smooth clutch and transmission operation; operable PTO and Motor Lift; good brakes, reasonable tires, and so on. Drive chains and sprockets should be checked for excessive wear or sloppiness.

In general, expect to pay more for a tractor on steel wheels.

R Series

In 1930, 60 percent of the nearly 6.3 million US farms were smaller than 100 acres. While farms under 100 acres constituted only 15 percent of all land under tillage, in number they represented a significant potential market to tractor manufacturers.

The general-purpose tractor offered the smaller farmer the same advantages as it did

The Model CO Vineyard Special featured 38in rear tread width, 10in narrower than that of the Model CO.

The Model CO orchard tractor. *Shields Library, Special Collections, University of California, Davis*

Rating	Model	Remarks
★★	R	Standard tractor; introduction followed that of RC by three years
★★★	RC	Featured Waukesha engine. Built in unstyled and styled versions; three- and four-speed transmissions; overhead and "chicken roost" steering

the larger farmer: a tractor that could operate in all crops and a ready source of belt-horsepower to power a variety of equipment.

However, a tractor the size of the Model CC was too large for many farmers. Follow-ing the lead of IH, who had introduced the Farmall F–12 in 1932, Case developed the Model RC, a one- to two-plow, row-crop tractor suited to the needs of the smaller farmer. Introduced in 1935 and built through

The Case Cane Special, Model CCS, featured 30in ground clearance.

The high-clearance Model CH, a modified standard Model C, featured 14in ground clearance at the front axle. Case built only seven units.

Early Model RC with overhead steering shaft and extended steering post. In 1937, the steering system was changed to the "chicken roost" style of the Model CC. *Case Company*

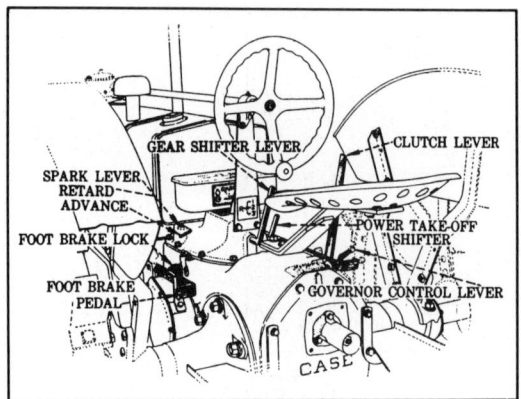

The Model RC operating controls were all within easy reach.

1936 Model RC Specifications
Recommended two-row cultivator, one-bottom 16in plow, or under favorable conditions a two-bottom 12in plow

Rated Brake Horsepower: 17
Rated Drawbar Horsepower: 11
Cylinders: Four: bore, 3¼in; stroke, 4in
Normal Engine Speed: 1425rpm
Ignition: Bosch MJB4 magneto with impulse coupling
Carburetor: Zenith
Fuel Capacity: Gasoline, 15gal
Cooling System: Capacity, 4½gal
Road Speed: 2⅓mph, 3⅓mph, and 4½mph
Total Weight: 3,350lb
Turning Radius: 7ft
Front Wheels: Diameter, 25in; width, 6in (single front, steel): 7: 50x10 (rubber)
Rear Wheels: Diameter, 48in; width, 2½in (narrow rim type, steel): 8: 25x36 (rubber)

1940, the RC featured a four-cylinder Waukesha engine with 3¼in bore and 4in stroke. Cast en bloc, it featured removable sleeves, drilled crankshaft, and oil bath air cleaner. Its centrifugal-type governor regulated engine speed at 1425rpm. Cooling was by thermosyphon, with a standard radiator and fan.

The RC's three-speed transmission and roller-chain final drive were similar to those employed in larger Case models and featured speeds of 2½, 3⅓, and 4½mph.

The rear tread was adjustable from 44in to 80in by the combination of sliding and reversing the rear wheels on the axle. Other features included standard PTO; independent differential brakes; optional fenders; optional extendable wide front axle, suited for use in bedded crops; choice of single or dual cast front wheels; and choice of steel or rubber tires.

The RC underwent a number of changes during its brief production run. Introduced with an overhead steering shaft and extended steering post, the system was changed to that used on the Model CC in 1937.

In 1939, the RC was restyled. A new cast-iron grille covered the previously exposed radiator. The tractor paint color was also changed from a light gray to the famous Case "Flambeau Red" (the color adopted for the new line of Case tractors introduced in 1939 and 1940).

Early Model RC on steel wheels.

Other changes in 1939 included the introduction of a standard four-speed transmission; Motor-Lift; and optional electric starter and lights.

The three-speed RC equipped with steel wheels was tested at Nebraska in April 1936 (Official Tractor Test No. 251). Its engine developed a maximum 19.80 brake hp and 13.26 drawbar hp. Maximum drawbar pull was measured at 2,103lb. In 1938, Case introduced the Model R, a standard tractor with fixed 46in tread. It featured the same Waukesha engine and three-speed transmission fitted to the RC. Fenders were standard; power takeoff and rubber tires optional. Industrial and orchard versions of the Model R were also produced.

R Series Production Totals by Model Year

Source: J. I. Case Company

1935	1936	1937	1938	1939	1940	Total
346	3500	2186	6200	2250	2903	17385

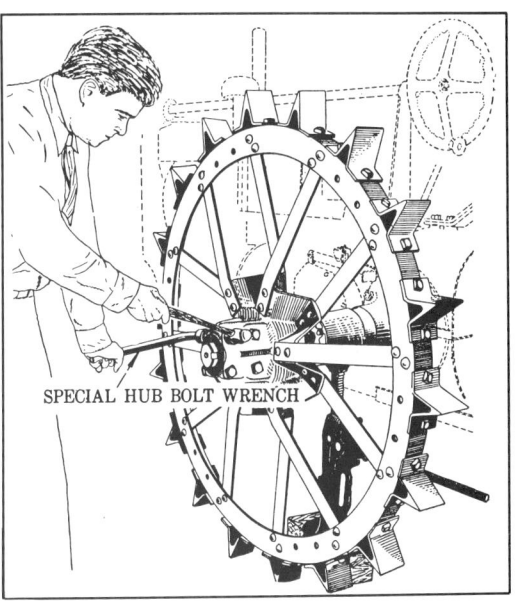

SPECIAL HUB BOLT WRENCH

Rear tread was adjusted by jacking up the Model RC, loosening the hub bolts with a special socket wrench that was supplied, moving the wheel to the desired position, and retightening the hub.

Early Model RC on rubber tires.

In 1939, the Model R was also restyled, painted Flambeau Red, and fitted with a four-speed transmission.

The R was tested at Nebraska in October 1938 (Official Tractor Test No. 308), equipped with rubber tires. The test unit developed maximum 20.52 brake hp and 15.58 drawbar hp. Maximum drawbar pull was measured at 2,574lb. Neither the R or RC were tested at Nebraska with the four-speed transmission fitted to the restyled tractors.

The Model RC was priced at $847 on rubber in 1940; the Model R at $655 on steel and $847 on rubber; and the Model RO (orchard tractor) at $892 on rubber.

Collecting Comments

R Series production was limited; only the V Series was built in fewer numbers. Its Waukesha engine and its governor were its weak points.

As collector tractors, interest in the Model R seems to be on the rise. While production was limited to a row-crop, standard, and orchard tractor, there is variety nonetheless. Consider the RC with overhead

Spacing range of extendable front axle varied from 56 to 80in, at 2in intervals; available with either steel wheels or rubber tires.

A Model RC with later style "chicken roost" steering.

The restyled Model RC arrived in 1939. The color was changed to Flambeau Red, and a four-speed transmission replaced the three-speed unit.

steering; chicken roost steering; unstyled design and three-speed transmission; styled (Flambeau version) and four-speed transmission; steel wheels; or rubber tires.

Expect to pay a premium for the early RC with overhead steering or an orchard version of the R.

The standard-tread Model R was not introduced until 1938. In 1939, it too was restyled and painted Flambeau Red. *Case Company*

Chapter 6

The Flambeau Red Tractors 1939–1955

The D, S, LA, V, and VA Series

Rating	Model	Remarks
★★★	D	1939 production and only if built with three-speed transmission. Total of 500 Ds and DCs began production as Cs, yet left the factory as Ds. All reportedly shipped to Canada
★★	D	The Model D standard tractor
★★★	DC–3	1939 production and only if built with three-speed transmission. Total of 500 Ds and DCs began production as Cs, yet left the factory as Ds. All reportedly shipped to Canada
★★	DC–3	The row-crop DC, single or dual front wheel
★★	C–4	Standard front axle, adjustable rear tread
★★	DH	This is the DC–4 built without turning brakes. Only 140 units built before production suspended
★★★	DO	Orchard tractor, but only with full fenders, branch deflectors, and cowling. Otherwise, no higher rating than that of standard Model D
★★★	DV	The vineyard special. Rates three stars by virtue of its limited production
★★★	DCS	The high-clearance sugar cane special. Also rates three stars by virtue of its limited production
★★	DI/STD	To many collectors industrial tractors have limited appeal, particularly to those who use their tractors for fieldwork or participate in plowing or tractor pulling contests
★★	DI/Narrow	Far more rare than the standard DI, which probably appeals only to hard-core Model DI collectors

The Flambeau Series of Case tractors was comprised of the D, LA, S, V, and VA Series, the bestselling tractors in Case history. During the Flambeau Red era, Case firmly established itself as the number three tractor manufacturer behind IH and Deere.

In retrospect, Case is sometimes criticized for its lack of innovation during this period. However, its models offered a range of horsepower matching that of its principal competitors and suited to meet the needs of a majority of farmers.

Case built nearly 390,000 tractors during this period, more than twice the number of Case gas tractors built prior to 1939. Consequently, regardless of potential investment

An early Model D, with optional rubber tires, pulling
a two-bottom plow. *Case Company*

An early Model DC equipped with optional rubber
tires. *Case Company*

1940 Model D Specifications

Pulls two, three, four plow bottoms—three bottoms under average soil conditions

Rated Brake Horsepower: 31.87*
Rated Drawbar Horsepower: 24.86*
Cylinders: Four: bore, 3⁷⁄₈in; stroke, 5¹⁄₂in
Normal Engine Speed: 1200rpm
Ignition: Case magneto with impulse coupling
Carburetor: Zenith 62AXJ9
Fuel Capacity: Gasoline, 17gal
Cooling System: Capacity, 6¹⁄₂gal
Road Speed: (On rubber) 2¹⁄₃mph, 3¹⁄₃mph, 4²⁄₃mph, and 9¹⁄₂mph
Total Weight: 7,005lb
Turning Radius: 10ft
Front Wheels: Diameter, 28in; width, 5in (single front, steel): 7: 50x16 (rubber)
Rear Wheels: Diameter, 42in; width, 11¹⁄₄in (steel); 12: 75x24 (rubber)
Height to Top of Hood (steel): 49¹⁄₂in
Wheelbase: 66³⁄₈in
Rear Wheel Tread: 48¹⁄₂in
Overall Width: 61¹⁄₂in
Overall Length: 116in
*Horsepower ratings from Nebraska Test results; tractor operated on distillate

Industrial Model DI Narrow Tread. Outside width measured 48in, versus DI Standard Tread range of 60 to 94.4in (equipped with dual wheels). *Case Company*

value, they are also the most commonly collected Case tractors.

D Series

The D Series replaced the C Series in 1939. It offered new styling and sported Flambeau Red paint but in most respects was simply an updated version of the C Series. Over its fifteen-year production run, Case made a number of changes and improvements, which kept the series competitive. When production ended in 1953, more than 100,000 units had been built.

The D Series was offered in several configurations, not all of which were built throughout the entire production run: standard-tread Model D; All-Purpose Models DC–3, DH, DC–4, and DC–4 Rice Special; orchard and vineyard tractors, Model DO and DV; both standard- and narrow-tread industrial tractors, Model DI Standard and Model DI Narrow Tread; a sugar cane special, Model DCS; and export versions of the Model D and DO. In addition, during World War II, Case built military versions of the DI.

Other than a change from a three-ring to four-ring piston, the D Series engine was virtually the same four-cylinder, 3⁷⁄₈in bore, and 5¹⁄₂in stroke unit fitted to the Model C. It operated at 1100rpm on the DC, and 1200rpm on the Model D.

Case offered a choice of manifolds on the D Series engine: the "many-fuel" manifold maintained higher manifold temperatures through an adjustable damper in the exhaust manifold and was designed to permit use of low-grade fuels; and the "economy gasoline" manifold, which Case claimed gave "full power and operating economy on any gasoline."

The first 500 units of the new D Series were built with the C Series three-speed transmission. Within the first few months of production, a four-speed transmission became standard. It offered a top speed of 9–10mph.

A Model DI with loader.

The Model DC–4 featured standard front axle and adjustable rear tread.

Standard Model D with rubber tires mounted on cut-down steel wheels, which was not at all an uncommon practice when the switch was made to rubber.

The standard Model D was tested at Nebraska in June 1940 (Official Tractor Test No. 349). The test tractor was equipped with rubber tires and the new four-speed transmission. Fueled by distillate, engine output was measured at 31.87 brake hp and 24.86 drawbar hp, under rated load.

The All-Purpose Model DC–3 was available with a choice of single front steel or cast-iron wheel; single rubber tire; or dual-wheel

D Series Production Totals By Model Year
Source: J. I. Case Company

Model	1939	1940	1941	1942	1943	1944	1945	1946*	1947	1948	1949	1950	1951	1952	1953	Total
DH	59	82	—	—	—	—	—	—	—	—	—	—	—	—	—	141
D	1162	1500	1128	740	112	1263	1	370	130	601	1044	2075	2000	1850	420	14396
DC–3	2564	2704	2098	1739	98	1496	2958	1595	5	5300	7580	9050	8714	5762	3622	55285
DC–4	143	238	404	458	385	680	570	—	751	1225	1525	2050	1937	1950	1700	14016
DO	66	219	296	—	—	—	—	220	1	376	406	430	575	285	—	2874
DO—VS	—	10	100	—	—	—	—	—	—	222	100	—	150	—	—	582
DCS	—	1	50	50	—	75	100	150	—	315	—	115	225	125	—	1206
DEX	—	710	372	346	351	817	105	410	355	1225	1049	400	450	650	—	7240
DOEX	—	—	5	—	—	50	20	25	—	—	50	20	25	15	—	210
DI/Std	—	83	254	663	948	1071	1364	250	2	1850	90	238	766	524	—	8103
DI/Nar	—	85	44	2	9	36	50	—	150	—	6	12	51	—	—	445

*Some confusion exists over production figures for the model year 1946. Case was hit by a labor strike in December 1945 that shutdown DC production for fifteen months. Yet, records show that 3,000 D Series units were built with serial numbers indicating 1946 production

1940 Model DC Specifications

Pulls two, three, four plow bottoms—three bottoms under average soil conditions

Rated Brake Horsepower: 32.94*
Rated Drawbar Horsepower: 25.74*
Cylinders: Four: bore, 3⅞in; stroke, 5½in
Normal Engine Speed: 1100rpm
Ignition: Case magneto with impulse coupling
Carburetor: Zenith 62AXJ9
Fuel Capacity: Gasoline, 17gal
Cooling System: Capacity, 6½gal
Road Speed: (On rubber) 2½mph, 3⅔mph, 5mph, and 10mph
Total Weight: 7,010lb
Front Wheels: Diameter, 25in; width, 5in (single front, steel)
Rear Wheels: Diameter, 48in (steel)
Height to Top of Hood: 54¼in (steel)
Wheelbase: 88⅞in (three-wheel)
Rear Wheel Tread: Adjustable, 48in to 84in
Overall Width: 55in to 91in
Overall Length: 134½in
*Horsepower ratings from Nebraska Test results; tractor operated on gasoline

tricycle front. The Model DC–4 featured a standard front axle, and choice of steel or rubber tires.

During the first two years of production, the rear tread width adjustment to the DC–3 and DC–4 was made by means of spools. The same combinations of 2, 10, and 12in spools and reversal of rear wheels, as used on the Model CC, permitted rear-tread adjustment from 48 to 84in. In 1941, the DC was fitted with longer rear axles and clamp-on wheels, which allowed tread adjustment through the combination of sliding and reversing the wheels on the axles.

The DC–3 and DC–4 featured standard power takeoff, Motor-Lift, and independent assisting brakes. (The Model DH, a DC with standard front axle that preceded the DC–4, did not offer turning brakes. Fewer than 150 units were built before production was suspended.)

Early Model DC–3 with optional electric starter and lights, pulling a Model A–6 combine. A power takeoff was standard on the Model DC and optional on the Model D. *Case Company*

Later Model DC–3 equipped with the Eagle Hitch. Note the clamp-on rear wheels, which, with a longer axle, replaced the extended hub or "spools" used for rear tread adjustment. *Shields Library, Special Collections, University of California, Davis*

Nicely restored Model DC–3 with an integral cultivator. Photographed at the WMSTR 1992 event.

1940 Model DO Specifications

Pulls two, three, four plow bottoms—three bottoms under average soil conditions

Rated Brake Horsepower: 31.87*
Rated Drawbar Horsepower: 24.86*
Cylinders: Four: bore, $3^{7}/_{8}$in; stroke, $5^{1}/_{2}$in
Normal Engine Speed: 1100rpm
Ignition: Case magneto with impulse coupling
Carburetor: Zenith 62AXJ9
Fuel Capacity: Gasoline, 17gal
Cooling System: Capacity, $6^{1}/_{2}$gal
Road Speed: (On rubber) $1^{3}/_{4}$mph, $3^{1}/_{3}$mph, $4^{2}/_{3}$mph, and $9^{1}/_{3}$mph
Turning Radius: 10ft
Front Wheels: 6: 00x16 (rubber)
Rear Wheels: 11: 25x24 (rubber)
Height to Top of Hood: $55^{3}/_{4}$in (rubber)
Wheelbase: $66^{3}/_{8}$in
Rear Wheel Tread: $50^{1}/_{2}$in
Overall Width: 62in
Overall Length: 116in
*Horsepower ratings as per Model D

The Model DC–3 was tested at Nebraska in April 1940 (Official Tractor Test No. 340). The test tractor was equipped with rubber tires and three-speed transmission. Fueled by gasoline, engine output was measured at 32.94 brake hp and 25.74 drawbar hp under rated load.

The D and DC had the following standard features: fenders, platform, swinging drawbar, belt pulley, radiator shutters and screen, and heat and oil gauges. Optional equipment included: starter and lights, PTO with guard (Model D), cushioned seat, muffler, a variety of grouter and lug styles for steel wheels, extension rims, rubber tires, and wheel weights.

The orchard tractor Model DO was offered with the same features of the Model D, plus independent brakes. Inside fenders and branch deflectors were regular equipment;

The early Model D featured a hand clutch mechanism on the right-hand side of the tractor; switched to the left hand side at serial number 4405303. In 1953, a foot-operated clutch was standard.

The right-hand side hand clutch mechanism of the early Model D was switched to the left-hand side at serial number 4405303.

A 1949 Model DC. A portable cylinder with hose lines and breakaway couplings was optional.

Shields Library, Special Collections, University of California, Davis

Proved CONVENIENCE

Many new Case tractor features are outstanding for their convenience as well as performance, economy or endurance . . . A long stroke engine with plenty of lugging ability and sensitive governor avoid unnecessary gears or gear shifting, quickly adapt your engine to all varying loads and conditions . . . A hot-spark magneto makes starting quicker, engine performance more positive . . . Large cooling system maintains uniform engine temperature for full power at all times . . . These are conveniences you do not readily see . . . But 50 modern conveniences that are easily found are described on the following page.

The Model DC operator platform and controls from a Case catalog. Hydraulic control, as shown, replaced the Motor Lift in 1949.

full outside skirted fenders, branch deflectors, and cowl were optional. The Model DV, a vineyard special based on the Model DO, differed only in its 10in narrower tread width.

The following changes and improvements were made to the D Series over fifteen years: a lowered operator's platform and relocated brake pedals, at serial number 4402475 (1940); a larger dash and instrument control panel, at serial number 4405303 (1940); a single-compartment fuel tank, fitted to tractors with the economy gasoline manifold; elimination of the hand brake and relocation of the clutch lever from the right-hand to left-hand side, also at serial number 4405303 (1940); a 112-tooth ring gear and new starter, at serial number 4404069 (1940); elimination of a separate operator platform, introduction of new seat bracket with spring suspension, and introduction of clamshell fenders (1941); introduction of single-disk brakes in place of the old band-type (1941); change in the gear shift pattern (1944); and a new clutch assembly (1944).

In the postwar years, Case introduced significant new features. In 1949, Hydraulic Control replaced Motor-Lift on the DC. A portable hydraulic cylinder added remote capabilities for pull-type implements. A heavy-duty clutch plate, high-capacity radiator, and heavier front spindle were also added in 1949.

In 1950, a high-compression head was offered and dual hydraulic controls were introduced. In 1951, the hood was redesigned, break-away hydraulic couplings were offered, and LP- Gas equipment was made available either as original equipment or for field conversion. In 1952, an adjustable wide front was offered for the Model DC–3; there was another change in platform; brake pedals were relocated from either side of the transmission housing to both pedals on the right side; dual-disc brakes were fitted; and, at serial number 5601106, live hydraulics and PTO and Eagle Hitch were offered (the Eagle Hitch was Case Company's own version of the three-point hitch developed by Harry Ferguson for Ford in 1939. First considered

Model "DO" Orchard and Grove Tractor.

In 1951, Case introduced LP-gas equipment for the D Series. Pictured is a Model DO Orchard and Grove Tractor.

Model DC–3 with optional, adjustable wide front.

A 1941 Model S. The early tractors are identified by their solid, flat rear wheels; deep shell fenders; intake and exhaust pipes that pass through the hood; chrome hood strip; and style and position of lights. *Case Company*

viable only for smaller tractors and implements, the three-point hitch was broadly accepted by the late 1940s. Case first offered Eagle Hitch on the Model VAC in 1949). During 1953, the last year of production, the hand clutch on models DC and D was replaced by a foot-operated clutch.

The Model DC–3 was priced at $1,040 on steel and $1,270 on rubber in 1944. The Model DC–3 was priced at $2,465 in 1952; the Model DC–4 at $2,182; LP-Gas option at $179; the Model DCS at $3,072; the Model DO at $2,168; and the Model DO Vineyard Special at $2,175.

Collecting Comments

The Case D Series is popular among Case collectors. It is not a particularly rare series of tractors. Sourcing parts is not a problem, which is terrific from the restoration standpoint. However, from an investment standpoint, the D is perhaps too common.

LP-Gas option was made available beginning in 1951. While some collectors may seek out an LPG unit, few would pay a premium for it. Assuming the system is operable, its major drawback is availability of fuel in comparison to gasoline or diesel. Yet, in my book, an LPG Model DO with full fenders, branch deflectors, and cowling borders on beautiful.

S Series

In 1941, Case introduced the S Series, a two-plow tractor positioned between the V Series and D Series. With the addition of the S Series, Case offered tractors in a range of hp that was a fair match to both IH and Deere.

The S Series was built in five distinct configurations: (1) the standard tractor Model S; (2) the row-crop Model SC, offered as a

A 1941 Model SC. The early SC engine block surface was smooth; later engines featured ribs cast in the block. *Case Company*

A Model SC–3 with extensible front axle, adjustable from 56 to 80in.

A Model SC–4 with standard front axle and adjustable rear tread was built for 1953 and 1954 model years only.

three-wheeler with choice of dual wheel tricycle front (in later units designated the Model SC–3), single front wheel for use in narrow rows, adjustable, wide front axle (Case used the term "extensible"); (3) the standard front axle, row-crop Model SC–4, produced in 1953 and 1954 only; (4) the grove and orchard tractor Model SO; and (5) an industrial tractor Model SI. Case also built export versions of the S and SO.

The S was promoted as a smaller version of the D. The general styling of the tractors was similar, as were the designs of their four-speed transmissions and roller chain final drives. The SC also featured the same "goose-neck" front wheel support and "chicken roost" steering arm as did the early

Model DC. However, the S Series engine clearly distinguished the tractor from its larger counterpart.

The S Series four-cylinder engine was a short-stroke, high-rpm engine designed to give maximum horsepower in a lightweight unit. At its introduction, the engine featured $3\frac{1}{2}$in bore, 4in stroke, and operated at 1550rpm. At serial number 8027115, during the 1953 production year, bore was increased to $3\frac{5}{8}$in. Stroke and engine speed remained unchanged.

Throughout production, Case offered gasoline-burning equipment as standard. Low-cost fuel equipment was optional. Such tractors were equipped with a small starting tank and radiator shutters.

The Model SC employed the same steering design and front wheel mounting of the early DC.

1942 Model S Specifications
Pulls two 14in bottoms under average conditions

Maximum Brake Horsepower: 21.62*
Rated Drawbar Horsepower: 16.18*
Cylinders: Four: bore, 3½in; stroke, 4in
Normal Engine Speed: 1550rpm
Ignition: Edison-Splitdorf magneto
Carburetor: Zenith
Fuel Capacity: Gasoline, 1¼gal; kerosene, 14gal
Cooling System: Capacity, 4gal
Road Speed: (On rubber) 2½mph, 3½mph, 4¼mph, and 10mph
Front Wheels: Diameter, 25in; width, 4in (steel): 5: 00x15 (4-ply rubber)
Rear Wheels: Diameter, 42in; width, 8in (steel): 10: 00x26 4-ply or 11x26 (4-ply rubber)
Height to Top of Hood: 51in
Wheelbase: 66in
Rear Wheel Tread: 46in
Overall Width: 56¼in
Overall Length: 108½in
*Horsepower ratings from Nebraska Test results; Model SC tractor operated on distillate

At introduction, the following features were standard equipment for the Model SC: oil bath air cleaner; deep-cushion, push-back seat; disc-type assisting brakes; adjustable drawbar; adjustable front and rear wheel tread; movable fenders, dash-panel engine temperature and oil-pressure gauges; belt pulley and power takeoff; and any combination of cast front wheels and narrow-rimmed steel rear wheels or front or rear rubber tires.

Standard equipment for the Model S was the same as that offered for the SC, with the exceptions of adjustable front and rear tread widths and power takeoff. The Model SO was equipped as the Model S and also featured full orchard fenders and cowl.

Special equipment for the early Model SC included the following: low-cost fuel equipment (as described above); starting and lighting equipment; Motor-Lift; muffler; a variety of weights, wheel equipment, lugs,

The Model SC ably handled two 14in plow bottoms.
Case Company

114

A 1952 Model SC with the Eagle Hitch. The broad spoke rear wheels were introduced in 1941. *Case Company*

and extension rims; radiator screen; and extra-length rear axle, which extended rear wheel tread from its standard range of 44 to 80in to a range of 44 to 96in.

1942 Model SC Specifications
Pulls two 14in bottoms under average conditions

Maximum Brake Horsepower: 21.62*
Rated Drawbar Horsepower: 16.18*
Cylinders: Four: bore, 3½in; stroke, 4in
Normal Engine Speed: 1550rpm
Ignition: Edison-Splitdorf magneto
Carburetor: Zenith
Fuel Capacity: Gasoline, 1¼gal; kerosene, 14gal
Cooling System: Capacity, 4gal
Road Speed: (On 10x38 rubber) 2½mph, 3½mph, 4¼mph, and 10mph
Total Weight: 4,200lb
Front Wheels: Diameter, 24¾in (steel): 5: 00x15 (4-ply rubber)
Rear Wheels: Diameter, 42in; width, 8in (steel): 10: 00x26 (4-ply rubber) or 11x26 (4-ply rubber)
Height to Top of Hood: 56in
Wheelbase: 82½in
Rear Wheel Tread: Adjustable, 44–80in
Overall Width: 74½in (minimum)
Overall Length: 126¼in
*Horsepower ratings from Nebraska Test results; tractor operated on distillate

The Case Eagle Hitch was introduced in 1949 on the VA. It was first offered on the S Series in 1952.

With the exceptions of Motor-Lift and extra-long rear axle and the inclusion of power takeoff, special equipment for the Models S and SO was the same as that offered for the SC.

The S Series remained in production into the 1954 model year (no units were built during 1946 due to a strike). During its production run the tractor underwent a number of modifications and improvements, which included: changes in hood design (1941); change from flat rear wheels to a six broad-spoke design (1941); relocation of the exhaust and intake stacks, which originally passed through the center of the hood, to the left hand side of the tractor (1941); shallower rear fender design (1942); hydraulic control (1950); Eagle Hitch (1952); and larger bore engine (as discussed above), foot-operated clutch and throttle, live PTO, dual valve hydraulic control, standard electric starter and lights (1953).

The Model S was priced at $845 on steel and $990 on rubber in 1944. The Model SO was priced at $1,322 on rubber in 1947; options included electric start, $49; PTO, $27; and Motor-Lift, $66.50. The Model SC was priced at $1,990 in 1952; the SC with adjustable front at $2,090.

Collecting Comments

The S Series did not sell as well as either the VA Series (approximately double the

Model SC Official Nebraska Tractor Test Comparison

Test Measure	Test No. 367* April 1941	Test No. 496** June 1953	Test No. 497*** June 1953
Max. Brake hp	21.62	29.68	23.67
Rated Drawbar hp	16.18	22.41	18.54
Rated Drawbar Pull	1,794lb	2,347lb	1,904lb
Max. Drawbar Pull	3,166lb	4,072lb	3,266lb

*3½x4in engine, low-cost fuel equipment (distillate)
**3⅝x4in engine, gasoline-fueled
***3⅝x4in engine, low-cost fuel equipment (tractor fuel)

S Series Production by Model Year

Model	1941	1942	1943	1944	1945	1946	1947	1948	1949	1950	1951	1952	1953	1954	Total
S	1314	100	855	719	450	—	800	700	926	1175	751	600	—	—	8390
SC	7540	5000	3722	4957	5330	—	3643	4438	7290	5526	5600	2900	1910	1135	58991
SO	246	300	—	—	150	—	249	221	311	130	135	75	—	—	1817
SI	—	100	1010	550	300	—	225	450	475	—	225	100	125	180	3740
S-Ex	154	100	84	214	185	—	840	391	—	909	100	300	350	—	3627
SO-Ex	4	—	—	—	—	—	1	—	40	10	15	—	—	—	70

number of S Series tractors) or the larger D Series. Nonetheless, the S Series was the third bestselling series of Case gasoline tractors.

From the collector's standpoint, however, little distinguishes the S. The orchard version is the most rare, followed by the earliest production units with flat wheels and the original hood design.

If the S appeals to you, the more popular features seem to be those of the later tractors, such as foot-operated clutch and live PTO.

Also, keep in mind that Model SC–4 production was limited. In the long run, it could well prove to be a more valuable configuration.

LA Series

Although the D Series replaced the C Series in 1939, the L Series remained unchanged and in production until 1940. Its replacement, the LA Series, was an updated version of the L, equipped with a new four-speed transmission, styled radiator grille and hood, and appropriate Flambeau Red paint.

One of several Model SC–3s featured at the 1992 WMSTR.

Rating	Model	Remarks
★★	LA	The standard-tread tractor
★★	LA/LPG	LPG-fueled standard-tread tractor
★★	LAI	Industrial tractor
★★★	LAIM	Military industrial version
★★★	LAH	Standard-tread tractor with Hesselman fuel-injected, diesel fuel-burning engine
★★★	LAIH	Industrial tractor with Hesselman engine

The LA Series included the standard Model LA, also offered as a rice special; and the industrial Model LAI, also built as a military tractor during World War II.

Available with either gasoline or low-cost fuel manifolds, the LA's four-cylinder engine maintained the $4\frac{5}{8}$in bore, 6in stroke, and 1100rpm operating speed of the L.

In 1942, Case introduced the standard Model LAH and industrial Model LAIH, which featured the Case engine modified to accommodate use of diesel fuels. The design,

Factory photo of the new Model LA. Electric starter and lights were optional at introduction, becoming standard in 1952. *Case Company*

Model LAI with standard electric starter, lights, and rubber tires. The front-mounted automatic coupler was optional. *Case Company*

known as the Hesselman engine, was not a true diesel. Rather, it was a fuel-injected, spark-ignited engine. By 1949, Case had built approximately 450 tractors with the Hesselman engine.

In 1952, the company introduced LP-Gas equipment for the LA. LPG- or butane-burning tractors were not uncommon in the 1950s and 1960s. Depending on location, LP-Gas was often much cheaper than gasoline. However, LP-Gas required special care in handling that kept most farmers from switching from gasoline.

The 1950s and 1960s were also the period in which diesel engines were gaining widespread acceptance. As diesel engines grew in popularity, the use of both gasoline and low-cost fuel alternatives dwindled.

The LA was not tested at Nebraska before August 1952, very near the end of its production run. Gasoline, distillate, and LPG versions of the tractor were all tested.

1952 Model LA Specifications*

Recommended four to five bottoms under ordinary conditions

Rated Brake Horsepower: 52.50
Rated Drawbar Horsepower: 41.30
Maximum Brake Horsepower: 58.53
Maximum Drawbar Horsepower: 51.68
Cylinders: Four; bore, $4^5/_8$in; stroke, 6in
Normal Engine Speed: 1100rpm
Ignition: Case magneto and electric starter
Carburetor: Vertical-type, with single nozzle; size, $1^1/_2$in
Fuel Capacity: $30^3/_4$gal
Cooling System: Capacity, $15^1/_4$gal
Road Speed: (On steel) $2^1/_4$mph, 3mph, 4mph, and $8^1/_4$mph; (on rubber) $2^1/_2$mph, $3^1/_3$mph, $4^1/_3$mph, and 10mph
Overall Length: 140in
Overall Width: $67^3/_4$in (steel); $75^1/_4$in (rubber)
Height to Top of Hood: $60^1/_2$in (steel); $61^7/_{16}$in (rubber)
Wheelbase: 82in
Weight: 7,621lb
Front Wheels: Diameter, 30in; width, 6in (steel); 7.50x18 (4-ply rubber)
Rear Wheels: Diameter, 48in; width, 12in (steel); 14x30 or 15x30 (rubber)
*Equipped with gasoline manifold

LA Series Nebraska Tractor Test August 1952

Test Measure	Gasoline	Distillate	LPG
Rated Brake hp	52.58	43.05	52.67
Rated Drawbar hp	41.63	34.99	40.90
Rated Drawbar Pull	3,347lb	2,769lb	3,274lb

Available options for the LA included the following: power takeoff, electric starting and lighting (standard in 1952); muffler (standard 1949); hood sides; front and rear extension rims; hydraulic control unit and portable ram attachment (1949–1952); yoke-type drawbar; low-low speed gear for rubber-tired tractors (1949–1952); independent hydraulic

LA Series Production Totals By Model Year
Source: J. I. Case Company

Model	1940	1941	1942	1943	1944	1945	1946	1947	1948	1949	1950	1951	1952	Total
LA	1203	1841	2180	593	3121	4037	—	1967	6023	7028	1900	2400	3200	35493
LAI	—	163	521	894	982	1407	—	58	1125	400	—	275	300	6125
LAH	—	—	10	12	17	58	—	—	125	200	—	—	—	422
LAIH	—	—	23	—	—	—	—	—	—	—	—	—	—	23

This military version of the LAI featured blackout lights, hood sides, muffler, starter, power takeoff, and winch. In 1943, Case delivered 450 units to the British Purchasing Commission under the Lend-Lease Act. An additional 200 units were built for US Army Ordnance. *Case Company*

Model LAH "Heavy Duty Oil Tractor" with the Hesselman system, fuel-injected, spark-ignition Case engine. The engine featured three-stage fuel filtration.

Case rated the Model LA as a four- to five-plow tractor. *Case Company*

rear wheel brakes (in 1951, disc-type differential brakes were introduced as standard equipment).

The Model L and LA reigned as the largest Case standard tractors for nearly a quarter of a century. Promoted for their four- to five-plow capacity, they were also the last of the heavy-duty, gasoline-powered standard Case tractors. Although the company continued to build standard gasoline tractors into the 1970s, the introduction of the Model 500 in 1953 marked the beginning of the diesel era.

The Model LA was priced at $1,360 on steel and $1,652 on rubber in 1944; electric starting was a $55 option; electric starting and lighting was a $72 option. The Model LA was priced at $3,030 (on rubber) in 1953.

Collecting Comments

The LA was the Case Company workhorse for thirteen years. By the early 1950s, it was dated (one collector I know refers to it as "clumsy"). Nevertheless, it was a very good, standard gasoline tractor.

The LA seems to have less appeal than the L, judging by the number of units that appear at various tractor events. However, there are rare variations of the Model LA to which many Case collectors are attracted, notably, the Hesselman LAH and LAIH, and the military versions of the LAI.

V Series

The V Series, built from 1940–1942, and the VA Series, built from 1942–1955, were the

MODEL "LA" CASE

In 1952, Case offered LP-gas equipment for the Model LA. Performance matched that of the gasoline-fueled engine, but it generally ran at a lower fuel cost.

Rear view of Model LA equipped with optional power takeoff.

Case promoted the Model LA's roomy platform and oversized seat that swung up for a "safety back rest" or could be pushed to one side, as preferred.

A line of Model LAs photographed at WMSTR 1992.

A Model VC tractor and Model F combine. The V Series was built largely from components purchased from outside vendors. *Case Company*

124

Rating	Model	Remarks
★ ½	V Series	Fitted with Continental engine and Clark transmission

smallest models in the Flambeau Series of Case tractors.

The VA was the bestselling Case tractor of the gasoline tractor era. It is to the Case collector what the John Deere Model B is to the Deere collector: a tractor available in number, with readily available spare parts, and easy to work on. However, its numbers limit its investment value.

The V was developed to meet the horsepower requirements of small farmers and commercial truck gardeners, for whom there were few alternatives other than draft animals. It is generally considered to have succeeded the R Series in its horsepower class.

The Model V four-cylinder engine produced 22.07 brake hp and 15.07 drawbar hp

The Model VC easily handled a plow with two 12in bottoms. *Case Company*

The industrial Model VI. Fewer than 750 units were built.

In the foreground is a Model VC with an optional starter and lights. In the background is a Model VC with an optional adjustable front axle.

under rated load. It was more tractor than offered by any of its major competitors. For example, its maximum drawbar pull of 2,798lb was nearly twice that of the popular Allis-Chalmers B.

The V Series was produced in four distinct versions: the fixed-tread, standard Model V; row-crop Model VC; industrial Model VI; and orchard Model VO. Of the 16,000 units built, 80 percent were the Model VC.

The V Series of tractors were built largely from components purchased from vendors. The valve-in-block engine, manufactured by Continental Motors, was the same unit used by Massey-Harris in the Model 101 R Junior. While Case cast the transmission and differential housings, the transmission and gears were supplied by Clark Company.

Rated as a one- to two-plow tractor, the V featured a four-speed transmission, foot clutch, and all-gear final drive. Power takeoff was standard on the Model VC; electric starter and lights were standard on the Model VI, but optional on the V, VC, and VO; an optional adjustable front axle was available for the Model VC, which was necessary for use in bedded crops. The standard rear axle on the Model VC permitted tread width adjustment from 44 to 82in. An optional axle package permitted narrower tread width adjustment from 36 to 72in.

Production of the V Series ended in October 1941, when it was replaced by the VA Series.

The Model V was priced at $605 on steel and $645 on rubber in 1940; the Model VC at $600 on steel and $630 on rubber.

Collecting Comments

The V Series was produced for a limited period of less than three years, before it was

A Model VAC, the standard row-crop configuration with dual wheel tricycle front, photographed at Rollag. Its distinctive "chicken roost" steering system was replaced in 1946 by a straight drag link.

V Series Production Totals by Model Year

Source: J. I. Case Company

Model	1940	1941	1942	Total
V	800	1111	410	2321
VC	2200	8160	2102	12462
VI	—	650	84	734
VO	—	324	187	511

I would not consider the V Series to be a particularly sought-after series of Case tractor. Nevertheless, expect to pay a premium for an orchard model, or a VC with optional narrow rear tread.

VA Series

By 1941, Case had developed its own engine to replace the Continental unit used in the V Series. The new engine was employed in the design of the VA, a new series of tractors that replaced the V Series early in the 1942 model year. Although it resembled the

replaced by the VA Series. It featured a Continental engine, which may make it less attractive to the Case enthusiast. It earned a reputation as a tractor that was hard to start, due to its poor magneto.

The standard tread Model VA was a versatile two-plow tractor.

Rating	Model	Remarks
★★	VA	Fixed front wheel tread
★★	VAC	Dual wheel tricycle front row-crop tractor
★★	VAC-11	Single front wheel row-crop tractor
★★	VAC-12	Same as VAC; 1951-1953 production
★★	VAC-13	Adjustable wide front row-crop tractor
★★	VAC-14	Adjustable wide front, low-profile row-crop tractor
★★	VAI	Industrial tractor with fixed front wheel tread
★★	VAO	Orchard tractor. Expect to pay a premium if equipped with orchard crown rear wheel fenders, rear fender closures, shield, and cowl
★★	VAO-15	Low-profile orchard tractor
★★★	VAIW	Warehouse and towmotor tractor. VAIW-3, single rear wheels; VAIW-4, dual rear wheels
★★★	VAH	High-clearance row-crop tractor with adjustable front and rear tread; narrow and wide axle
★★★	VAS	Offset high-clearance row-crop model

A 1952 production VAC with the then-standard starter and lights. *Case Company*

**1942, 1944–1953 Model VA
Specifications**
Standard tractor

Wheelbase: 75.25in
Front Wheel Tread: 43in (early units); 48.25in (late units)
Rear Wheel Tread: 44–72in adjustable (early units); 48–72in (late units with Eagle Hitch)
Options: PTO; belt pulley; Eagle Hitch (from 1949); electric start (standard from 1948); Low Cost Fuel (from 1947)

V, the VA shared no major components with its predecessor.

VA production began in November 1941 and, except for production of military units, was interrupted in July 1942 due to World War II. The entire industry faced shortages, as both material and labor were redirected toward wartime manufacture. Production of farm tractors resumed in 1944.

Sales of the VA proved strong, as the war came to an end and demand for tractors pushed industry sales to record levels. In December 1945 (1946 model year), a strike at Case closed all plants except those at Rock Island and Burlington, Iowa. The strike lasted fifteen months, during which time the VA was the only tractor Case built. As a consequence, its sales mushroomed.

Production of the VA peaked in 1948, after which it steadily declined. In 1953, production was halted by another labor dispute. After several months, production resumed on a strictly limited basis into the 1956

The Model VAH high-clearance tractor offered 26³/₄in clearance; adjustable front wheel tread of 61 to 85in; and adjustable rear tread of 60 to 88in.

Shields Library, Special Collections, University of California, Davis

model year. Despite the interruptions of war and labor strife, nearly 150,000 VA Series tractors were built during its fourteen-year production run.

The VA Series included the following models: the standard-tread Model VA; row-crop Model VAC (available with single front wheel, dual wheel tricycle front, or adjustable front axle); industrial Model VAI; warehouse and military tow motor tractors, Model VAIW, VAIW–3, VAIW–4; orchard tractors, Model VAO and Model VAO–15; high-clearance Model VAH; and offset, high-clearance, Model VAS.

The VA four-cylinder engine featured bore of $3\frac{1}{4}$in and stroke of $3\frac{3}{4}$in, valve-in-head design, three-bearing crankshaft, and removable cylinder sleeves. It operated at

General dimensions of the Model VAO Orchard and Grove Tractor were identical to those of the standard Model VA. Without orchard shield and cowling or rear fender closures, the absence of upright intake and exhaust are the most obvious clues to aid in identification of this 1952 Model VAO. *Case Company*

131

Model VAC–14 with low-profile platform, seat, and steering. Built the final two years of VA production, it featured adjustable front and rear axles, standard starter, and lights.

VA Series Production by Model Year

Model	1942	1943	1944	1945	1946	1947	1948	1949	1950	1951	1952	1953	1954	1955	Total
VA	1740	—	947	648	1715	2350	3717	3227	1128	1413	561	137	—	—	17583
VAC	8500	988	4834	6458	13027	10367	13992	11973	6831	10215	6193	889	—	—	94267
VAC–14	—	—	—	—	—	—	—	—	—	—	—	4467	1759	—	6226
VAI	320	1630	647	1050	1944	1854	2934	570	707	837	805	649	521	563	15031
VAO	440	—	—	175	216	1817	1547	567	237	382	347	196	265	736	6925
VAIW	—	—	744	2860	44	32	—	—	—	80	23	10	—	—	3793
VAH	—	—	—	—	—	221	445	562	68	238	194	45	68	175	2016
VAS	—	—	—	—	—	—	—	—	—	18	1028	374	139	—	1559

1425rpm. For the first several years, the engine was built by Continental Motors with castings supplied by Case. In June 1947, Case assumed complete production at its newly opened Rock Island engine plant.

Earliest VA units operated on gasoline only. In 1947, Case introduced the Low Cost Fuel (LCF) system, which permitted use of distillate fuels. The LCF system had the following features: a special manifold with heat control valve; separate fuel tanks (the smaller tank or "starting tank" for gasoline) and three-way fuel shut-off valve; radiator shutter; and a thermostat that remained

Side view of a Model VAC with adjustable front. Throughout most of VAC production, front and rear lights were optional.

closed until engine temperatures reached a level to effectively burn the low-grade fuel. Beginning in 1949, Case also offered the engine with special high-altitude pistons.

Tractors with both gasoline and distillate were tested at Nebraska in October 1949. In Official Tractor Test No. 430, a VAC equipped with LCF developed 15.93 brake hp and 12.54 drawbar hp under rate load. Maximum drawbar pull was measured at 2,394lb. In Official

Tractor Test No. 431, a VAC equipped with the standard gasoline engine developed 19.04 brake hp and 15.00 drawbar hp under rated load. Maximum drawbar pull was measured at 2,768lb.

The VA Series transmission featured four forward speeds in a range from $2\frac{1}{8}$ to 12mph, depending on model and wheel equipment. Unlike the rest of the Case line, the VA employed an all-gear final drive.

A number of notable changes, additions, and modifications were made to the VA Series over the course of its production. The

Rear view of a Model VAC.

original steering system on the VAC employed the "chicken roost" steering arm, as first used on the Model CC. This was replaced in 1946 by a straight drag link tucked in close to the tractor. In 1951, a new steering design was introduced that employed universal joints and enclosed worm and gear on the front spindle.

1951–1953 Model VAC-12 Specifications
General-purpose tractor

Wheelbase: 84in
Front Wheel Tread: dual front wheel, adjustable 6.5–12.5in
Rear Wheel Tread: 48–88in
Standard: Eagle Hitch; hydraulic lift; electric start
Options: PTO; belt pulley; Low Cost Fuel

1951–1953 Model VAC-13 Specifications
General-purpose tractor

Wheelbase: 77in
Front Wheel Tread: adjustable wide front 52–76in (standard); 46–68in and 44–60in (optional)
Rear Wheel Tread: 48–88in
Standard: Eagle Hitch; hydraulic lift; electric start
Options: PTO; belt pulley with idler pulley attachment; Low Cost Fuel

1953–1954 Model VAC-14 Specifications
General-purpose tractor, identical to VAC-13 but with lowered operator platform and shortened steering column

Wheelbase: 77in
Front Wheel Tread: adjustable wide front 52–76in (standard); 46–68in and 44–60in (optional)
Rear Wheel Tread: 48–88in
Standard: Eagle Hitch; hydraulic lift; electric start
Options: PTO; belt pulley with idler pulley attachment; Low Cost Fuel

1942–1955 Model VAI Specifications
Industrial tractor

Wheelbase: 75.25in
Front Wheel Tread: 43in (early units); 48.25in (late units)
Rear Wheel Tread: 44–72in adjustable (early units); 48–72in (late units with Eagle Hitch)
Standard: Adjustable tilt-back, upholstered seat
Options: PTO; belt pulley; Eagle Hitch with hydraulic lift (from 1949); electric starter and lights (standard from 1948)

The VA Series engine, designed by Case, was built by Continental Motors before Case opened its Rock Island engine plant in 1947.

Built during the model years 1951–1954, the Model VAS High Clearance tractor was designed as a one-row cultivator for use in tall, bushy crops such as tomatoes or asparagus; for peppers, celery, and other truck crops; and for nursery trees and shrubs.

The Model VAIW–3 industrial tractor was built to travel through narrow aisles and make sharp turns in restricted areas. In low gear and on concrete flooring, it developed 3,000lb drawbar pull.

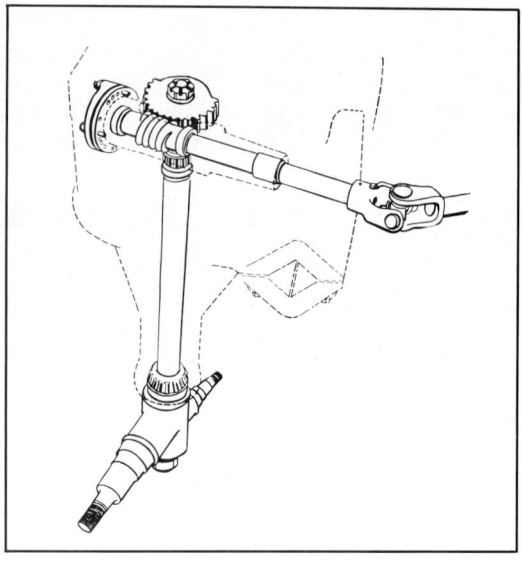

In 1951, the Model VA steering system was modified to incorporate a universal joint and enclosed worm and gear on the front spindle.

The VA Series was first to employ the Case Eagle Hitch and hydraulic lift, introduced in 1949. The system was standard on the VAC, VAH, and VAS; optional on the VA, VAO, and VAI.

In 1951, Case introduced new model designations for the single front wheel, dual front wheel, and adjustable front axle versions of the VAC already in production; the VAC–11, VAC–12, and VAC–13 respectively.

In 1953, Case introduced the Model VAC–14. Similar to the VAC–13, it featured a lowered platform, lowered seat, shortened steering column, adjustable wide front, longer rear axle, and it offered electric starter and lights as standard equipment.

The orchard and vineyard tractor, Model VAO–15, was also introduced in 1953. Based on the existing VAO, it featured a lowered platform and seat, shortened steering column, and adjustable front and rear wheel

Sectional view shows the simple 4-speed transmission, gear shift control cover and the centrally located power take-off shaft.

The Case-built Model VA four-speed transmission and all-gear final drive.

A Model VAC–13 with adjustable front axle. Note the U-joint in the steering link. This unit equipped with Eagle Hitch is pulling a Model 280 border disc plow. *Case Company*

treads. It also offered Eagle Hitch and electric starter and lights as standard equipment.

The Model VA was priced at $710 on rubber in 1942; electric starter and distributor was a $32 option. The Model VAO was priced at $1,113 in 1948. The Model VA was priced at $1,388 with standard electric start and $1,538 with hydraulic lift in 1952; the Model VAO was priced at $1,388; the Model VAH at $1,766; the Model VAC–11 at $1,535; the Model VAC–12 at $1,499; the Model VAS at $1,564; the Model VAC–13 at $1,577 with Eagle Hitch, electric starter, and fenders; distillate manifold option was priced at $43.50. The Model VAO–15 was priced at $1,611 in 1955 with Eagle Hitch and PTO; distillate manifold option was priced at $44; and belt pulley at $43.

Collecting Comments

The VA Series is to the Case collector what the John Deere Model B is to the Deere collector. A tractor that almost every farmer who farmed with Case tractors in the forties, fifties, and sixties owned at some time. Many units remain at work as cultivating tractors, all-purpose tractors, or mowing tractors for small plot holders and weekend farmers. When I was a Case dealer in the late seventies and early eighties, it was the type of tractor we rarely took in trade but could resell in a flash.

The VA is a wonderful series of tractors that offers a variety of advantages to collectors. VAs are plentiful, available in a variety of

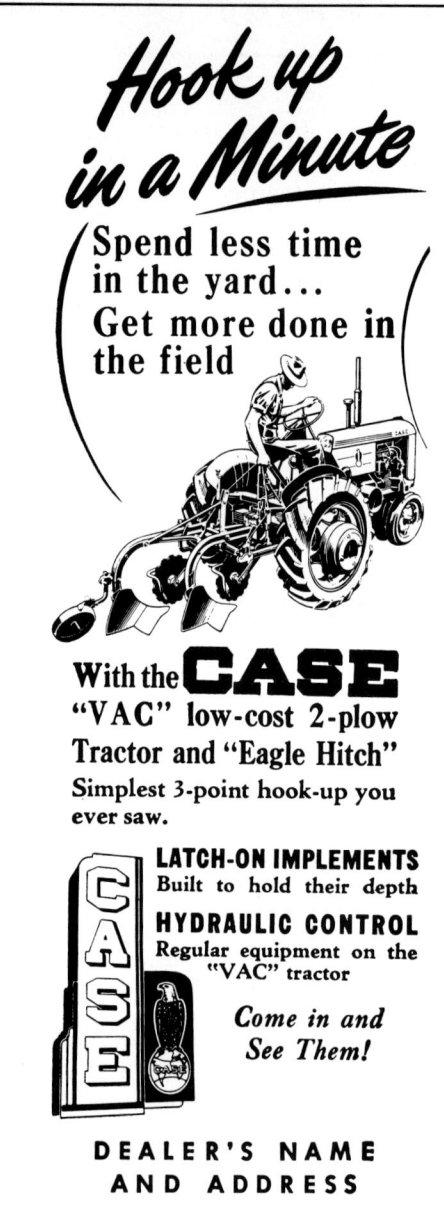

Case Eagle Hitch ad.

1942–1947, 1051–1953 Model VAIW Specifications
Warehouse and towmotor tractor: VAIW 3, single rear wheel military unit; VAIW-4, dual wheel unit

Specifications: 1800rpm; four-speed transmission; 3,000lb drawbar pull; electric start, disk brakes, front bumper, rear coupler

---

1942, 1945–1955 Model VAO Specifications
Orchard and vineyard tractor

Wheelbase: 75.25in
Front Wheel Tread: 43in (early units); 48.25in (late units)
Rear Wheel Tread: 44–72in (early units); 48–72in (late units with shell fenders); 52in with crown fenders
Standard: shell type fenders; cushioned seat
Options: Orchard crown rear wheel fenders; orchard shield and cowl attachments; rear fender closures; PTO; belt pulley; Eagle Hitch with hydraulic lift (from 1949); electric starter (standard from 1948) and lights (standard from 1953); Low Cost Fuel (from 1947)

1947–1955 Model VAH Specifications
High-clearance general-purpose tractor

Wheelbase: 84in
Front Wheel Tread: 61–85in (standard); 53–73in (narrow)
Rear Wheel Tread: 60–88in (standard); 52–64in (narrow)
Crop Clearance: 27in
Standard: Eagle Hitch with hydraulic lift (from 1949); electric starter (standard from 1948) and lights (standard from 1953)
Options: PTO; belt pulley and idler attachment; Low Cost Fuel

1951–1954 Model VAS Specifications
Offset high-clearance general-purpose tractor

Wheelbase: 69.5in
Front Wheel Tread: 42–58in
Rear Wheel Tread: 42–70in
Crop Clearance: 24.25in under front axle; 23.75 under rear axle
Standard: Eagle Hitch with hydraulic lift; electric starter and lights
Options: PTO; belt pulley and idler attachment; Low Cost Fuel

configurations, easy to work on, easy to find parts for, easy to transport, easy to store, and easy to enjoy.

However, the fact that it is so common makes the VA Series less attractive as an investment. Here is where sentiment and heart come into play. While there is no Case tractor that I would rate with only one star, enthusiasts considering the VA should keep in mind that it would be easy to invest more in restoring one than could be recouped at resale time.

In conclusion, I would say the VA is an ideal first tractor for the Case enthusiast or for someone who would enjoy a do-it-yourself restoration project.

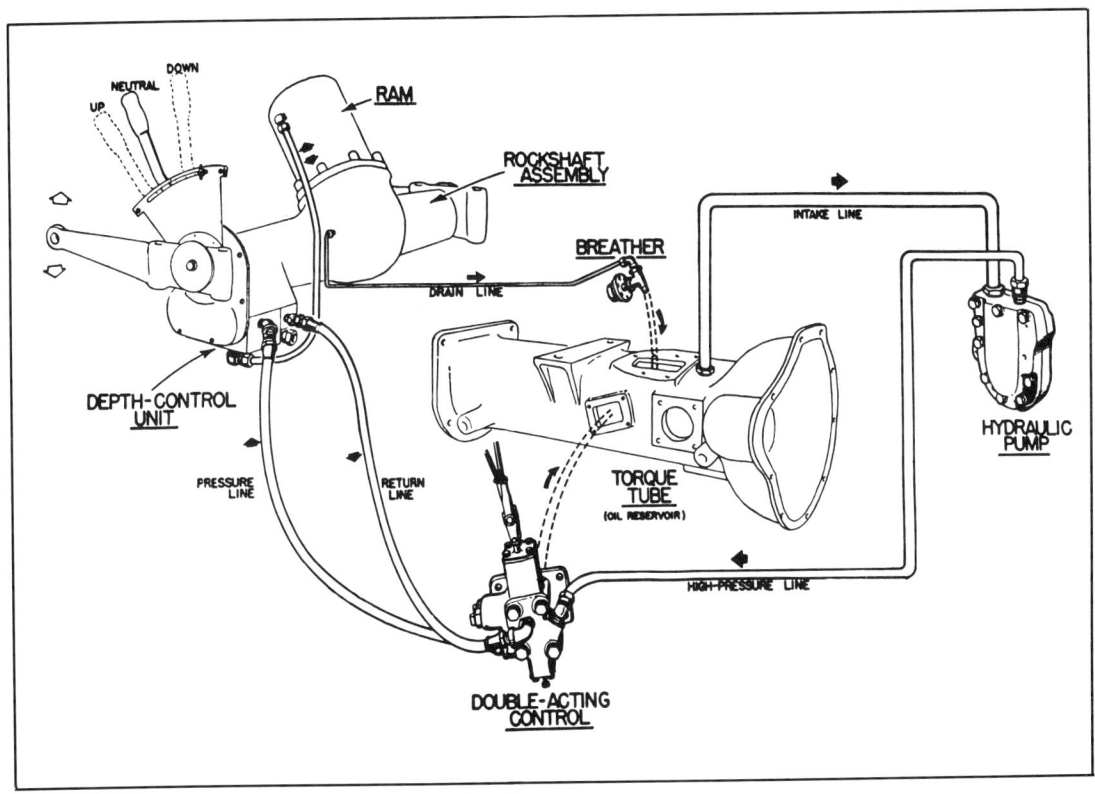

A schematic of the Eagle Hitch and hydraulic system, as fitted to the Model VAI.

Here is a still-employed, hard-working Model VAC–
14 with mower deck. Note the low-profile platform
and steering.

NEW CASE TRACTOR

VAO-15

for

ORCHARDS and VINEYARDS

The Model VAO with lowered platform and steering was transformed into the Model VAO–15, in 1953.

VA Series Total Production

Model	Years	Total
VA	1942, 1944–1953	17,583
VAC	1942–1950	94,267*
VAC-11	1951–1953	**
VAC-12	1951–1953	**
VAC-13	1951–1953	**
VAC-14	1953–1954	6,226
VAI	1942–1955	15,031
VAIW	1942–1947, 1951–1953	3,793
VAO	1942, 1945–1955	6,927***
VAH	1947–1955	2,016
VAS	1951–1954	1,559

* Includes Models VAC-11, VAC-12, and VAC-13
** Included in Model VAC
*** Includes VAO-15

Case Gas Tractor Serial Numbers

Case Gas Tractor Serial Numbers

Changes in a particular model's design, components, or parts were often made during the course of its production. The changes, and the serial number at which such changes occurred, are generally noted in parts and service manuals. Consequently, serial numbers are a significant bit of information, valued by the tractor owner.

Serial numbers are also a guide to the year of manufacture. Many collectors place a premium on early production units or on blocks of units built with specific characteristics. In many cases serial numbers distinguish such units.

Case used more than one method of numbering tractors. Beginning in 1912, serial numbers were assigned in blocks, but without regard to year.

Case Serial Numbers 1912–1928

Year	Serial Number
1912	100 to 890
1913	891 to 2495
1914	2496 to 2841
1915	2842 to 3690
1916	3691 to 7491
1917	7492 to 13284
1918	13285 to 22222
1919	22223 to 32840
1920	32841 to 42255
1921	42256 to 43942
1922	43943 to 45280
1923	45281 to 48412
1924	48413 to 51677
1925	51678 to 55918
1926	55919 to 62408
1927	62409 to 68403
1928	69004 to 69803

In 1928, Case introduced a six-digit serial number that can be decoded in two steps to determine the model year. In the first step, the first and fourth digit are combined (not added together) to create a two-digit number. For example, for the serial number 344196, combine 3 and 1 to create 31. In the second step, subtract 3 from the number created in step one. In our example, 31 minus 3 equals 28. Using this method, it is determined that serial number 344196 was built for the model year 1928.

In 1938, Case switched to a seven-digit serial number. For most tractors built through the 1953 model year, the year of manufacture can be determined by a similar two-step method. In the first step, the first and second digit are combined. For example, for the serial number 4205587, combine 4 and 2 to create 42. In the second step, subtract 4 from the number created in step one. In our example, 42 minus 4 equals 38. Using this method, you can determined that serial number 4205587 was built for the model year 1938.

Beginning in 1954 serial numbers were assigned in blocks. For the models included in this guide, the beginning serial number for each year follows:

VA Series

Year	Serial Number
1953	5750001
1954	6011001
1955	6038001

D and LA Series

Year	Serial Number
1953	5750001
1954	5800001

S Series

Year	Serial Number
1953	5700001
1954	8035001

Case Gas Tractor Paint Colors

Paint color for Case tractors varied by era. The colors used and supplier references, as of November 1992, are as follows:

Dark Green: Used on early crossmotor tractors through 1922 or 1923. Sources: NAPA 99L 8748; Ditzler 40249.

Dark Red: Used on wheels and front axles of crossmotor tractors, L and C Series. Sources: Case B13010; Dupont 066DH; Ditzler 73252.

LC Gray: This is the darker shade of gray used on crossmotor models from mid–1923 through 1929, all L and C Series. Source: Case B13017.

RC Gray: This is the lighter shade of gray used on the R Series, prior to its restyling in 1939. Sources: NAPA 99L 11464; Dupont 87310; Ditzler DTR 32137.

Flambeau Red: Used on restyled R Series, D, S, LA, V, and VA Series. Sources: NAPA 99L 3727; Dupont 016DH; Ditzler 71282; PPG 70138 or PPG 71282.

Trim, Wheels, and Accessories

According to *Old Abe's News,* radiator lettering was silver, hood lettering and wheels dark red on all late crossmotor models and all L, C, and CC tractors; wheel rims were painted silver or aluminum on Flambeau models after 1941, black on earlier tractors; starter, generator, lights, muffler, exhaust manifold, and distributor were painted black on Flambeau models; intake manifolds were painted Flambeau red on Flambeau models; and magneto color varied by vendor.

Index